Individual Resilience to Urban Flooding and the Implications for Urban Management

Jing Song

Individual Resilience to Urban Flooding and the Implications for Urban Management

Springer

Jing Song
School of Urban Planning and Design
Peking University
Shenzhen, Guangdong, China

ISBN 978-981-97-0545-0 ISBN 978-981-97-0546-7 (eBook)
https://doi.org/10.1007/978-981-97-0546-7

This Springer imprint is published by the registered company Springer Nature Singapore Pte Ltd.
The registered company address is: 152 Beach Road, #21-01/04 Gateway East, Singapore 189721, Singapore

Paper in this product is recyclable.

Contents

Abbreviations

DRR	Disaster Risk Reduction
E-B	Environment-Behavior
Gongming	Gongming sub-district in Shenzhen, China
HLM	Hierarchical linear model
PE	Personal efficacy
PEOF	Previous experiences of urban flooding
PEP	Physical environment perceptions
PRP	Personal risk perception
SCP	Social cohesion perception
SSP	Social support perception
SWMM	Storm water management model

List of Figures

List of Tables

Chapter 1
Introduction

The adverse impacts of climate change associated with rapid urbanization, such as rising sea levels, prolonged precipitation, and frequent and intense rainfall, are emerging as primary threats to our earth (White, 2013). Since urban areas are the center of economic and political activities and home to over half of the world's population (Miao et al., 2011), heavily populated areas, particularly low-lying coastal cities, are vulnerable to flooding caused by these events (Lo et al., 2015). Despite the massive development of flood-control infrastructures such as channelization, dams and levees, flooding remains a dangerous, unpredictable, and frequent disturbance threatening cities worldwide (Shao and Xu, 2015).

City systems are dynamic, self-organizing systems (Ahern, 2013; Allen, 2012; Pattee, 2012) that can withstand disturbances of a certain level. However, with the rising influence of human behavior on urban systems, such urban systems nowadays are more complex and changeable than before. Developing measures that can expand the capacity of urban systems to withstand and adapt to a wider array of external shocks and disturbances has become an urgent concern (Bino et al., 2014; Desaraju & Michael, 2014; Kalnay & Cai, 2003; Ziska et al., 2003). This in essence concerns the concept of resilience, which is widely accepted as a metaphor, theory, set of capacities and measure dealing with such flux-uncertain and unpredictable state (Davoudi, 2012). Understanding this concept and its implications on urban sustainability has thus become an important area of research (Ahern, 2011; Holling, 1973, 1986; Leichenko, 2011).

The traditional approach to flood management emphasizes engineering resilience, which primarily focuses on the engineering design of infrastructures against urban floods (Booth, 1991; Nickel et al., 2014). However, with the increasingly unpredictable, diverse and intense disturbances affecting cities, it appears that a pure physical approach to tackle urban flooding is inadequate (Benson et al., 2007; Shao & Xu, 2015; Tran & Shaw, 2007; Xu & Ou, 2014). For instance, among the 8 casualties caused by Typhoon Hato's attack in Macau, 2 people died because going

J. Song, *Individual Resilience to Urban Flooding and the Implications for Urban Management*, https://doi.org/10.1007/978-981-97-0546-7_1

into underground carparks to ‘save’ their cars during severe urban floods.[1] In addition to the inadequacy of physical protection, these incidents also raise doubts over the capability of residents to protect themselves during severe natural disasters. It is evident that besides “hard” measures, advancements in “soft” measures from a social perspective to enhance resilience should also be on the agenda of the government and other stakeholders in urban management.

The concept of individual resilience is considered here a complex concept determined by both internal (individual) and external (community) attributes, which focuses on the social aspect of resilience. At the community level, the quality of physical design and layout of individual communities, as well as the social environment, could play important roles in building individual resilience (Ahern, 2011; Smith, 2014). At the individual level, personal perception and awareness to natural disasters could be a major factor affecting one’s ability to withstand urban flooding (Chamlee-Wright & Storr, 2009; Cohen et al., 2016). Nonetheless, one should be aware that while individuals are regarded as the single unit of behavior, which collectively determine the overall resilience of a community, their perceptions to their living communities, e.g. physical and social environment, also influence the resilience of individuals (Cox & Perry, 2011; Silver & Grek-Martin, 2015). Therefore, it is necessary to explore the dynamics created by different forms of interactions, i.e. the complex “people-disaster-place-society” relationship influencing one’s individual resilience.

Effective urban management would be key to advancing in these aspects, shaping the built environment and the quality of communities (Fekete et al., 2014; Quarantelli, 1988; Turner, 1994). For instance, through comprehensive housing programs and urban planning, especially high-order strategic planning, neighborhood quality, such as built environment, crime rates and social atmosphere, could be enhanced (Jiang et al., 2013; Pahl-Wostl, 2007). For instance, communities planned under national-level programs, such as the Qianhai Pilot Free Trade Zone in Shenzhen, China, could be endowed with considerably more resources and expertise to enhance neighborhood quality. Supervision from urban management for residential estates, could make the day-to-day management of residential estates be more organized and comprehensive through regular programs for hardware maintenance as well, contributing to the integrity of the social and physical environment of the community (Norio et al., 2011).

Given the significance of urban management in the built environment and quality of communities, it should have significant influence on individual resilience considering the complex interactions among people, place, society and disaster. Therefore, urban management should be a top concern for resilience building. According to the adaptive cycle theory, social learning could result in substantial improvements in urban management (Armitage, 2006; Smith & Stirling, 2010; Tran & Shaw, 2007). Through the acquisition of experience from previous disasters and advancements in knowledge base, the competence of different stakeholders in urban management,

[1] http://time.com/4917981/typhoon-hato-macau-aftermath/.

including central and local governments, informal organizations, industries and residents, could be enhanced. Therefore, it is necessary to analyze the problems in the current social learning and explore the dynamics between urban management and individual resilience, in turn lifting the resilience of individuals.

1.1 Background: Urban Flooding and Resilience in China

Developing countries are most affected by abrupt climate change. Floods, as a natural disaster resulted from climate change, have affected many areas worldwide, particularly developing countries, resulting in economic damage and human misery cities (Mirza, 2003). In China, the increase in private car use, intensification of environmental pollution, and similar trends associated with rapid urbanization have led to huge carbon dioxide emission, in turn causing climate change and more frequent droughts and heavy rain (Feyen et al., 2009; Konrad, 2003; White & Greer, 2006). In addition, as a developing country, the large amount of construction projects has also led to dramatic changes in land use (Tanner et al., 2009). The occupation of a large extent of natural land has led to continuous reduction in the amount of natural water, fragmentation of the natural waterway network, and an increasingly impervious surface, making the urban system more vulnerable to intensive precipitation and thus severe urban flooding (Huong & Pathirana, 2013).

Data from the Ministry of Housing and Urban–Rural Development of the People's Republic of China revealed that of the 351 cities investigated nationwide from 2008 to 2010, 62% were affected by flooding.[2] 137 of them were even flooded for more than three times within three years. Worth noting is that major urban centers such as Beijing, Shanghai, Guangzhou, and Shenzhen, are also in the list of cities prone to serious flooding caused by heavy rain.[3] In 2012, urban flooding of Beijing, China killed 78 people and caused an economic loss of CNY 11.64 billion to the capital city of Beijing,[4] again bringing this threat to the attention of the government, researchers and different stakeholders. Even worse, the old, poorly designed and under-maintained sewage system in China has greatly reduced its efficiency in handling floodwaters (Xie, 2013). It can be said that in addition to the environmental uncertainties, the rather reckless approach to city development may have caused the high vulnerability to urban flooding.

Researchers and the government are constantly finding measures to combat urban flooding. However, revamping the sewage systems in densely built-up areas, if not impossible, would require massive inputs in terms of time, money and energy. In 2013, the Chinese government developed a nationwide agenda of Sponge City Initiative,[5] envisioning building cities that are capable of resisting severe urban flooding by

---

[2] http://cpc.people.com.cn/GB/64093/82429/83083/15176547.html.

[3] https://www.turenscape.com/news/detail/1660.html.

[4] https://en.wikipedia.org/wiki/July_2012_Beijing_flood.

[5] http://www.mohurd.gov.cn/zxydt/201411/t20141102_219452.html.

absorbing and storing excessive rainwater and releasing it for use during droughts. In particular, the measure of Low Impact Development (LID),[6] a stormwater management measure that uses microscale and localized practices to control the runoff caused by storms and maintains the pre-development hydrological characteristics of the development site (SZDB/Z 145-2015), was adopted. With a spending of CNY 86,500 million, the Chinese government aims to realize this vision to reduce urban flooding and alleviate the problem of droughts within three years.[7] Although the concept of resilience is not explicitly touched upon in official documents on Sponge City Initiative, this measure is in essence an attempt to build resilience to heavy rain and flooding. Nonetheless, despite the bold vision and substantial recourses devoted, there have been doubts as to whether this national-level measure could be implemented effectively given the very diverse local contexts in different communities across the nation. Some also raised concerns over the excessively strong focus on the physical aspect of Sponge City Initiative, paying insufficient attention to the social dimension in building resilience.

An illustrative example of communities severely affected by floods is Gongming Sub-district in Guangming New District, Shenzhen, China (Gongming). Existing news[8] suggest it is a place vulnerable to urban flooding. Demographically, it is home to many peasant workers with around 80% of its dwellings being urban villages. Therefore, Gongming is an ideal case study for examining the limited individual resilience of the poor and vulnerable peasant workers. With its designation as the pilot area of the Sponge City Initiative in 2015, it also offers an opportunity for examining the effectiveness of this initiative in improving individual resilience. The exploration of place-responsive management measures based on its unique location and social-economic characteristics of its population would be valuable as well.

1.2 Research Objectives, Questions and Hypotheses

This study explores urban management measures to enhance the performance of individual resilience. First, the performance of individual resilience is explored, forming the basis for identifying the limiting factors of individual resilience. At this stage, individual resilience is framed as the capacity of coping with and adaptation to urban flooding. Self-reported coping and adaptive strategies adopted by individuals as well as their communities' that can empower individuals are taken as the basis for evaluating individual resilience performance. Then the variations in individual resilience performance among residents of different communities are explored. Second, the influencing and limiting factors of the resilience to urban flooding are identified and explained at both the individual and collective levels. Further, based on the limiting factors identified, the institutional setting and implementation program of

[6] http://www.mohurd.gov.cn/zxydt/201411/t20141102_219452.html.

[7] http://www.mohurd.gov.cn/zxydt/201605/t20160510_227414.html.

[8] http://www.szmb.gov.cn/site/szmb/zhongdianquyu/info/2017/32356.html.

the nationwide Sponge City Initiative is evaluated to ascertain if it can improve individual resilience performance level for different social groups. Finally, place-responsive recommendations to enhance individual resilience are provided based on the previous analyses.

In line with these objectives, there are three research questions provided based on three hypotheses:

(1) What was the current performance of individual resilience for the residents vulnerable to urban flooding? And did the individual resilience vary among different communities?
(2) What were the determinants (People-Disaster-Place-Society) of individual resilience?
(3) Whether the strategic measure of Sponge City Initiative could help nurture individual resilience? And what were the place-responsive measures for individual resilience nurturing?

Hypotheses are made to the aforementioned questions:

(1) There are significant differences in individual resilience to urban flooding for residents in different communities.
(2) Individual perceptions of urban flooding, living place and living society can influence individual resilience.
(3) Individual resilience cannot be nurtured through resilience-oriented measure, i.e. Sponge City Initiate from central government of China.

1.3 Significance

Conventional research on the concept of resilience to urban flooding is primarily focused on the community level, giving limited attention to other levels such as the individual, household, city and national scales. Taking individuals as the analytical unit to identify the resilience performance characteristics of different social groups, this research deepens the understanding of the concept of resilience at the individual level. A particular innovation in this research is that it extends the understanding of individual resilience to a People-Disaster-Place-Society interaction. Not only does this open the possibility to use individual behavior to evaluate individual resilience, but to explore the personal perceptions and cognitions as determinants of resilience performance is also a breakthrough in studying the concept of resilience.

This study differentiates itself from existing literature in that it undertakes a major quantitative assessment on the concept of individual resilience. By calculating individual resilience based on coping and adaptive strategies to urban flooding, the constraints on individual resilience in terms of personal perceptions as well as the variances of communities are quantitatively identified. Disaster risk deduction is also embodied in the analysis of building resilience performance, bringing these into a unified conceptual framework for understanding the resilience of cities and how they should move towards a more resilient state.

In addition, while resilience is a concept widely studied in the West, there has been a lack of research on this domain in China. Through a comprehensive analysis of resilience to urban flooding in the Chinese context, more in-depth and dynamic comparison between the performance of urban resilience in China and the West is made possible. A strong basis for further research on resilience in China is also provided.

Finally, this research frames individual resilience and its potential relationship with place-responsive urban management based on social learning theory. By evaluating the effectiveness of general management measures in nurturing individual resilience, this research provides insights on the significance and directions for formulating place-responsive measures, supplementing the literature on the tactics of nurturing resilience for poor communities with high vulnerabilities to urban flooding.

1.4 Structure of Book

The following chapters of this book give detailed account of the issues mentioned above. Chapter 2 gives an introduction to the concept of resilience, presenting the definition and domain of individual resilience and its relationships with adaptation and coping strategies, individual perceptions, and disaster management. A conceptual framework developed for this research based on literature review is also presented. Chapter 3 describes the threats brought by and responses to urban flooding at a global scale and national scale in China. Particular focus is given to the Shenzhen, the case city for this research. In Chap. 4, a complete analytical framework is explained based on the theoretical link among the elements associated with the research hypotheses and questions. Besides, the research data, methods and case area selection, i.e. Gongming in Shenzhen, China are detailed introduced and explained. Chapters 5, 6, and 7 present the study on individual resilience, its influencing and limiting factors, and measures for resilience improvement. Chapter 8 summarizes the research findings, policy implications, and the limitations and contributions of this study, identifying directions for future research.

Chapter 2
Conceptual Background: Resilience and Urban Management

2.1 Introduction

To comprehend the improvement of individual resilience in urban management in China, this chapter lays the theoretical foundation of the study by reviewing the literature on resilience and urban management. Resilience theory is explained as the foundation of this research. The theory of human–environment interactions is adopted to explain why the personal perceptions and cognitions can be explored as potential determinants of individual resilience. Furthermore, social learning for resilience nurturing is adopted to explain why and how the responses from urban management can make people more resilient.

This chapter comprises four sections. The first section presents the concepts and theories of resilience. Based on the review, we summarize the research objective: individual resilience. Capacities of individual resilience and their relationship with human behavior are summarized. Based on relationship, the first research hypothesis of this study is formulated. The second section summarizes the relationships between individual resilience and personal perceptions of disasters and living environment. Based on these relationships, the second research hypothesis of this study is formulated. The third section reviews the theoretical foundation of interventions from urban management to facilitate attaining the desired individual resilience. Based on this, the third research hypothesis is formulated. Furthermore, gaps in the research are identified, and attention turns to the conceptual framework in the fourth section.

J. Song, *Individual Resilience to Urban Flooding and the Implications for Urban Management*, https://doi.org/10.1007/978-981-97-0546-7_2

2.2 Concept of Resilience to Natural Disturbances

2.2.1 Natural Disturbances and Resilience Thinking

Today, we live in a world threatened by disturbances associated with catastrophic climate change (Davoudi, 2012). We are entering an era of change that calls for rapid and flexible responses (Berkes & Ross, 2013). A disturbance upsets the normal state, changing the structure, dynamics, productivity, and biodiversity of urban systems (Calderon-Aguilera et al., 2012). In the past, this external force stemmed from nature; now, human activities such as dramatic land use changes, substantial carbon dioxide emissions, and deforestation contribute to these events (Chen et al., 2014; Steward, 2012). As such, natural disturbances are threatening the economic development of cities and human physical and mental health (Carmin et al., 2009; Leichenko, 2011; IPCC, 2007; Willems et al., 2012). Simultaneously, resilience thinking (resilience) focuses on adapting to this state of flux from the global to local level of management (Davoudi, 2012; Holling & Gunderson, 2002).

2.2.2 The Realm of Resilience

Resilience involves an exchange of thought across disciplinary borders, such as among biology, engineering, geography, economics, and catastrophology (Keck & Sakdapolrak, 2013; Windle, 2011). Nowadays, resilience is also widely accepted as the theoretical and practical methodology to mitigate natural disturbances (Ahern, 2013; Cote & Nightingale, 2012; Martin, 2012; Plieninger & Bieling, 2012; Walklate & Mythen, 2014; Wieland & Marcus Wallenburg, 2013).

As a field of interdisciplinary research, many scholars have offered their own definitions on resilience to disturbances (Table 2.1).

Based on Table 2.1, the development of the definition of resilience is briefly summarized.

Holling (1973) defined resilience from a physical perspective. Holling's definition of resilience is the ability of a system to return to an equilibrium or steady-state after a disturbance (Holling, 1973, 1978, 1986). The speed for a system to come back to its equilibrium state can be adopted to measure resilience. The faster the system bounces back, the higher its resilience. Therefore, analysis of the return time is the key, and "efficiency, constancy and predictability" is the pursuit of this type of research (Holling, 1996). It can be concluded that this definition focuses on resistance to disturbances and depends on the effectiveness of physical infrastructures and modern technologies. This is referred to as engineering resilience or physical resilience, which can be adapted to research on engineering design (Holling, 1996).

Later, the definition was broadened to encompass structural changes (Gunderson, 2000). Ecological resilience is provided as the "amount of disturbance that an

Table 2.1 Definitions of resilience

Author	Definition
Wildavsky (1988)	Resilience is the capacity to cope with unanticipated dangers after they have become manifest, learning to bounce back (p. 77)
Holling et al. (1995)	It is the magnitude of disturbance that can be absorbed before the system changes its structure by changing the variables and processes that control behaviour (p. 50)
Mallak (1998)	For your organization to be resilient, it needs people who can respond quickly and effectively to change while enduring minimal stress (p. 1)
Dunning (1999)	Resilience describes an active process of self-righting, learned resourcefulness and growth-the ability to function psychologically at a level far greater than expected given the individual's capacities and previous experiences
Kendra and Wachtendorf (2003)	The ability to respond to singular or unique events (p. 42)
Pelling (2003)	Resilience to natural hazard is the ability of an actor to cope with or adapt to hazard stress (p. 48)
Resilience Alliance (2005) (http://www.resalliance.org/576.php)	Ecosystem resilience is the capacity of an ecosystem to tolerate disturbance without collapsing into a qualitatively different state that is controlled by a different set of processes. A resilient ecosystem can withstand shocks and rebuild itself when necessary. Resilience in social systems has the added capacity of humans to anticipate and plan for the future
Perrings (2006)	The ability of the system to withstand either market or environmental shocks without losing the capacity to allocate resources efficiently (the functionality of the market and supporting institutions), or to deliver essential services (the functionality of the production system (p. 418)
Haimes et al. (2008)	Resilience as the ability of the system to withstand a major disruption within acceptable degradation parameters and to recover within an acceptable time and composite costs and risks (p. 291)
Aven (2011)	Resilience as the uncertainty about and severity of the consequences of the activity given the occurrence of any types of events (p. 518)

Sources Adapted from Joseph (2013), Liao (2012), Manyena (2006), and Wang et al. (2012)

ecosystem could withstand without changing self-organized processes and structures (Gunderson, 2000, p. 427)." It is "the magnitude of disturbance that can be absorbed before the system redefines its structure by changing the variables and processes that control behavior (Gunderson, 2000, p. 426)." Therefore, resilience does not necessarily imply a return to the pre-existing state of the ecosystem. In contrast, resilience can be viewed as the capacity to respond to these changes and assume another stable status (Holling, 2001). This distinguishes ecological resilience from engineering resilience.

Resilience now focuses on adaptation to disturbances (Adger et al., 2005; Davoudi, 2012). Although systems can recover and reorganize after disturbances, they do not have to return to their original status (Adger et al., 2005; Berkes et al., 2002; Davoudi, 2012). Socio-ecological or evolutionary resilience is provided as the capacity of social-ecological systems to absorb recurrent disturbances such as hurricanes or floods so as to retain essential structures, processes, and feedback (Berkes, 2007). According to this definition, resilience is the capacity to absorb, self-organize, learn, and adapt to complex disturbances (Cosens et al., 2013; Folke, 2006; Walker et al., 2002, 2006). From this perspective, Folke (2006) interpreted resilience as the set of "opportunities that disturbance opens up in terms of the recombination of evolved structures and processes, renewal of the system, and emergence of new trajectories (pp. 259)." In this context, resilience is an adaptive cycle that allows for continuous development in a dynamic interaction between stability and change, involving complex interactions between society and ecosystems.

2.2.3 *Adaptive Cycle Theory*

2.2.3.1 System and Panarchy Theories

The evolution of resilience research has produced the research realm of socio-ecological resilience, based on Holling and Gunderson's (2002) modern resilience theory of adaptive cycle theory.

Adaptive cycle theory emphasizes that the most important feature of socio-ecological resilience is a dynamic loop. Systems theory is used to describe a status in which everything is related to everything else within the complex systems (Simon, 1977). Complex systems include the functional, compositional, and structural hierarchies (Hong & Espelage, 2012). The linkage among them is demonstrated by their functions, which in turn produce additional compositional and structural elements (Dale & Beyeler, 2001). Socio-ecological systems are such integrated with societies and environments (Berkes & Turner, 2006; Forbes et al., 2009; González et al., 2008). It does not mean the simple summation between social systems and ecological systems, while the interaction between the two systems. Therefore, the socio-ecological system, also known as the human-natural system, focuses on the human behavior working on nature (Berkes & Turner, 2006; Forbes et al., 2009). The urban system is just an example of the socio-ecological system, with characteristics of non-deterministic and dynamic, and is easily influenced by a multitude of economic, social, physical, and institutional factors (Ahern, 2013; Carpenter et al., 2005; Jabareen, 2013; Paton, 2006; Ungar, 2012). To conclude, socio-ecological resilience is multidimensional and changeable with human behavior working on natural system.

Panarchy is another theoretical model to describe the complex system as inherently dynamically organized and structured across scales of space and time (Allen et al., 2014; Cumming et al., 2006). Rather than pursuing the outdated equilibrium paradigm, panarchy theory explains the paradoxes within the multi-equilibrium context, for examples, connected versus adaptable, flexible versus efficient, resilient versus transformational, and persistence versus change (Davoudi, 2012; Gunderson, 2000). Panarchy is not only a simple hierarchical structure demonstrated through top-down and larger-scale processes, but also bottom-up processes occurring on a small scale. In other words, panarchy theory supports the viewpoint that processes within complex systems have "cross-scale linkages, therefore, processes at one scale can affect those at other scales to influence the overall dynamics of the system" (Allen et al., 2014; Redman & Kinzig, 2003).

2.2.3.2 Adaptive Cycle and Its Loops

An adaptive cycle with an infinity loop is shown in Fig. 2.1, which was developed based on systems and panarchy theories. The adaptive cycle does not emphasize a single equilibrium point any more, while the multiple metastable regimes (Folke et al., 2002). Therefore, the adaptive cycle is a dynamic process across different scales of space and time. Resilience theory tries to explain the source and role of changes from the perspective of adaptive cycles, which can be graphically represented by an infinity loop (∞) encompassing four phases: growth/exploitation, conservation, release/creative destruction, and re-organization (Folke et al., 2002; Moore & Westley, 2011; Redman & Kinzig, 2003). The first phase (exploration, r) is the beginning of the cycle, and relates to the present structures and functions of the complex socio-ecological system (Holling & Gunderson, 2002; Pendall et al., 2010). The second phase (conservation, K) is a relatively long period that cumulates and absorbs disturbances. If the disturbances have not exceeded the maximum absorption capacity, it is locked into this period. However, it is inevitably followed by the third phrase (release, Ω) when self-organization is urgently needed (Davoudi, 2012; Holling & Gunderson, 2002; Walker et al., 2006). The fourth phase (reorganization, α) is a period during which previous systems collapse and systems learn to innovate and find new paths for re-organization and renewal, either slowly or rapidly (Davoudi, 2012; Holling & Gunderson, 2002; Martin, 2012; Pendall et al., 2010; Simmie & Martin, 2010). Phase Ω and phase α comprise an unpredictable "backloop," and "the α phase leads into a subsequent r phase, which may resemble the previous r phase or be significantly different" (Walker et al., 2004).

The following conclusions are based on the theoretical model. First, the four phases are not necessarily sequential or fixed. Second, the status transformation can be a series of interactive nested circles. Third, there is a constantly changing process, not a fixed asset, end, or outcome. Finally, the process combines different dimensions of resilience in terms of the dynamic interplay of persistence, adaptability and transformability across multiple scales and timeframes.

Based on the evolutionary concept of socio-ecological resilience and its inherent theory-adaptive cycle, in terms of natural disturbances, it can be concluded that resilience to natural disturbances (hereinafter referred to as "resilience") is a loop, and can only be realized and observed when there is a natural disturbance. Moreover, resilience changes towards the desired resilience through the learning in the fourth phase of the adaptive cycle. The desired resilience means the research object has the desired level of the ability and capacity to cope with and adapt to disturbances (Bruneau et al., 2003; Walker et al., 2004). Further resilience occurs at different levels (individual and collective) and before, during, and after a disturbance.

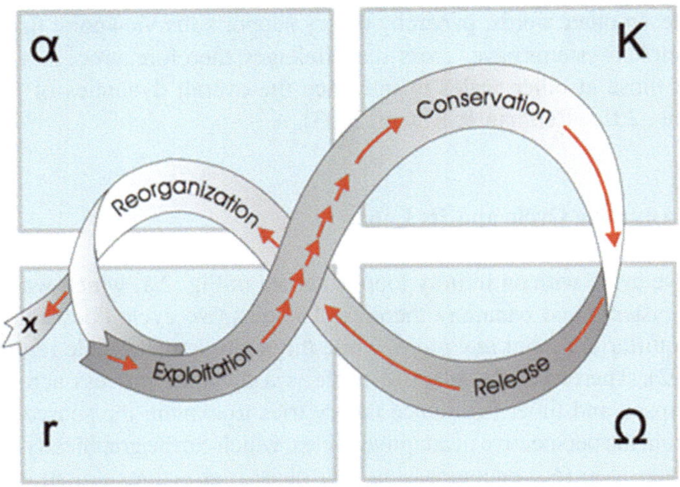

Fig. 2.1 The panarchy model of the adaptive cycle. *Source* Adapted from Holling and Gunderson (2002, pp. 34–41) and Pendall et al. (2010, p. 76)

2.2.4 Framing Individual Resilience

2.2.4.1 Dimensions of Resilience

Socio-ecological resilience does not only consider the physical resilience of system, but focusing on its social dimension as well as the interactions between the social and physical systems.

Physical resilience focuses on the built and natural environments such as buildings, roads, infrastructure, facilities communications, and geographic factors such as the soil, terrain, and topography (Ahern, 2011; Smith, 2014). During an external disturbance, the ecosystem ensures an individual or a community's functioning and survival. However, if the disturbance exceeds the physical resilience capacity of the ecosystem, then the city is vulnerable (Godschalk, 2003; Khailani & Perera, 2013) to the disturbance.

Social resilience is another aspect of resilient residents or communities (Maguire & Hagan, 2007), covering the "formal and informal, stable and ad hoc human associations that operate in an urban area, such as schools, neighborhoods, agencies, organizations, enterprises, task forces, and the like (Hirokawa, 2014, p. 155)." Human beings are responsible for responding to disturbances and learning from past experiences. If the responders of disturbances have quite low capacities without external supports empowering them, the city is even more vulnerable (Goldstein, 2012; McIlwaine & Datta, 2014).

We are now living in a world of unpredictable natural disturbances. The traditional strategies to curtail urban flooding such as enhancing the infrastructure cannot envisage an impregnable city that can withstand floods. Therefore, only focusing on physical protection cannot ensure resilience to natural disturbances (Brown & Westaway, 2011). As the victims of urban flooding, social units are exposed to multiple disturbances. Internationally, efforts are being made to reduce social vulnerability (Cutter et al., 2008). Therefore, social resilience, especially social entities-individuals, social groups, communities and their capacities to absorb, cope with and adapt to disturbances, which show the interaction between social and physical systems, deserve research (Keck & Sakdapolrak, 2013).

2.2.4.2 Individual Resilience

Emphasizing social resilience draws attention to the quality of life of the social entities who are vulnerable to disturbances and to develop strategies to enhance their original resilience capacity and realize a better outcome when the disturbance comes (Glavovic et al., 2002; Obrist et al., 2010). The social entities consist of layers of resilience, which can be summarized as individual resilience, community resilience, and societal resilience (Obrist et al., 2010).

In urban systems, community resilience is an important component of resilience that links individual and societal resilience (D'Odorico et al., 2010; Plough et al., 2013). Nonetheless, in the literature on community resilience we can conclude that (1) residents of communities are the focused components of community resilience (Chandra et al., 2010); (2) the responses and preparedness of individuals to disturbances within a community should be the key elements of resilience (Folke et al., 2003; Norris et al., 2008; Plough et al., 2013); (3) the community as the research unit, however measuring community resilience should be based on individual or household resilience (Beatley & Newman, 2013; Jabareen, 2013; Walsh, 2015). All these suggest that although community resilience is the mainstream of resilience research worldwide, individual resilience deserves attention as it informs the ways in which people cope with and adapt to disturbances.

2.2.4.3 Capacities of Resilience and Individual Resilience

The United Nations Office for Disaster Risk Reduction (DRR) (2002) accepts that resilience is a set of capacities of an individual, community, or social systems to fight against a natural disturbance to maintain or realize an acceptable level of functioning and structure.[1] To conclude, there are three capacities (absorption capacity, coping capacity and adaptive capacity), three temporal scales (pre-event during event and post-event/post-disaster), and two spatial scales (individual and collective) of resilience. In the context of social resilience, coping and adaptive capacities of resilience are introduced detailed in the following part.

Adaptive capacity. Adaptive capacity/adaptation is closely related to resilience, and some scholars even equate them (Gallopín, 2003). With the development of

[1] https://www.unisdr.org/we/campaign/cities.

resilience theory, adaptive capacity is accepted as an important component of resilience (Carpenter & Gunderson, 2001; Gunderson, 2000), and another important component of resilience is its coping capacity (Henly-Shepard et al., 2015; Lorenz, 2013; Yohe & Tol, 2002).

The literature points out that adaptive capacity is strategic capacity and requires the stakeholders to have the ability to anticipate, plan and take reflexive actions when facing natural disturbances (Brown & Westaway, 2011; Hill & Engle, 2013; Saldaña-Zorrilla, 2008). Gunderson (2000) has remarked that adaptive capacity depends on people's ability to learn and adapt as reflected in their behavior. It is the capacities of actors, including individuals and social groups, to respond to, create and even to shape the disturbances (Chapin et al., 2009; Walker et al., 2004). Brown and Westaway (2011) define adaptive capacity as "the preconditions necessary to enable adaptation to take place, it is a latent characteristic that must be activated to effect adaptation." To conclude, adaptive capacity is one of the important components of resilience and it is the capacity and ability enabling the engagement of all stakeholders to fight and adapt to change instead of returning to the status quo (Berkes et al., 2000; Maguire & Cartwright, 2008; Vugrin & Camphouse, 2011). It focuses on human actions, their willingness to undertake adaptation, the resources to be adopted and deployed, as well as the constraints upon out undertaking adaptation. The stakeholders that demonstrate adaptive capacity have different spatial scales, ranging from individual to collective (Ensor et al., 2015). It can take place on different temporal scales, but is demonstrated as strategic capacity, which is particularly important after a disturbance.

Coping capacity. Coping capacity is reflected when it is assumed that a disturbance happens in a familiar pattern, therefore the previous coping actions can be worked as the guide for the similar events (Béné et al., 2015; Gaillard, 2010; Norris et al., 2008). It is the capacity of individuals or groups to use their resources and their abilities to cope with the disturbances to their livelihoods and needs (Jabeen et al., 2010). While adaptive capacity is under "the scenario of a changing climate, risks may increase but adaptation actually expands a system's coping ranges" (Jabeen et al., 2010). Besides, the functional temporal scales between adaptive and coping capacity are different (Cutter et al., 2008; Folke, 2006). The coping capacity of individuals or groups plays a dominant role during and immediately after a disturbance, and it requires immediate action (Cutter et al., 2008; Jabeen et al., 2010; Norris et al., 2008; Parsons, 2016). In contrast, adaptive capacity is especially important after a disturbance, and it can influence and change the coping capacity of individuals and groups over the long term (Abramson et al., 2015; Folke et al., 2002; Parsons, 2016).

Coping and self-organization. Self-organization includes independent agents whose individual responses, in aggregate, lead to a systemic response to external disturbances (Few, 2003; Turner, 2014). The capacity for self-organization in collective behavior is the fundamental reallocation of energy and action to achieve a resilience goal in the event of a disturbance. Self-organization is developed based on social networks, and "a network is a self-organized and typically informal governance system in which diverse actors are knitted together across organizational levels to focus on common problems (Folke et al., 2005, p. 450)." Therefore, on the individual level, coping capacity can be viewed as the individual's ability to handle disturbances. On the collective scale, self-organization can be viewed as the ability of social groups to handle disturbances.

Adaptation and transformation. The literature states that adaptation is a process of resilience. Adaptation consists of actions undertaken by individuals, and an expression of adaptive capacity (Adger & Vincent, 2005; Brown & Westaway, 2011). Transformation is a capacity of government, and depends on its governance, infrastructure and service delivery, formal and informal social protection mechanisms, as well as the policies/regulations that constitute the conditions for unexpected disturbances (Folke et al., 2002; Plummer & Armitage, 2010). It should be interpreted as the collective and strategic actions supporting by formal and informal organizations, and usually works when the original urban systems have collapsed after serious disaster, such as after serious earthquake (Folke et al., 2002, 2010; Plummer & Armitage, 2010; Wilson et al., 2013). On the individual level, adaptive capacity can be viewed as the adaptation of individual to adapt to *natural disturbances*, while on the collective scale, transformative capacity can be viewed as the transformation of social groups to adapt to *natural disasters*, which are supported by the actions of governments.

Combining the definitions of the evolution of resilience to disturbances, the adaptive cycle theory, capacities, social resilience can be conceptually framed as two timeframes:

(1) During and immediately after the disturbance period: The degree to which socio-ecological systems are capable of coping (individual level) or self-organization (collective level) when it is still raining (versus lack of organization or organization forced by external factors), which is at the social scale.
(2) A long period after and before the next disturbance: The degree to which the socio-ecological system can increase its capacity for adaptation (individual level) or transformation (collective level) after the disturbance, at the interactively physical and social scales (Fig. 2.2).

These two aspects of the concept are uninterrupted succession or flow, a coherent whole, and occurrences that follow one another when combatting disturbance, which forms the resilience loop (Fig. 2.2). The theoretical definition and realm of resilience have been framed to form the theoretical foundation of this research.

In terms of individual resilience, we can further narrow down capacities of resilience into individual level to urban flooding:

(1) The degree to which the individuals are capable of coping heavy rain and urban flooding when it is still raining; and (2) The degree to which the individuals can adapt to urban flooding through learning and innovation after the urban flooding disasters.

2.2.5 *Individual Resilience and Human Behavior*

Adaptive capacity theory points out that in complex socio-ecological systems, human behavior interacts with the external disturbances associated with climate change.

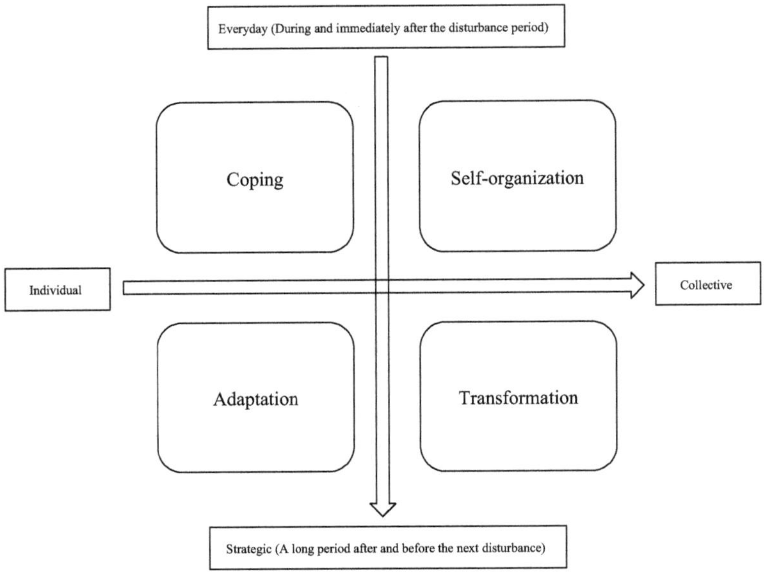

Fig. 2.2 A conceptual framework for resilience to natural disturbances. *Source* Author revised based on the research of Brown and Westaway (2011)

This has led scholars to believe that individual behavior can reflect their capacity for resilience. The "hit-damage-recovery-learning cycle" theory (Lei et al., 2014) shows how human behavior reacts to disturbances and demonstrates their ability to cope, recovery and adapt, i.e. resilience capacity. In order to calculate individual resilience capacity, we should identify the behavior of people to react to urban flooding, and then judge if the behavior exhibits resilience thinking (Berkes, 2007; Legaspi et al., 2014). Understanding human behavior towards the socio-ecological systems they live in requires knowing not only how people act today, but also what they think about the future (Goldman & Riosmena, 2013; Peterson, 2000). Therefore, it is identified that in evaluating individual resilience performance, it is the coping and adaptive behavior adopted by the individuals as well as the collective behavior which can support, serve and facilitate the individual behavior towards desired resilience that should be focused on (Adger, 2003; Christmann & Aw-Hassan, 2015). Theory of action involves corporate actors at the societal level and persons at individual level (Coleman, 1986; Harper, 2015). Facing disturbance, human behavior demonstrating the capacity for coping and adaptation does not only include the behavior adopted by individuals, but the one adopted by corporate actors, which can empower individual behavior (Harper, 2015). Such behaviors are as follows. First is the knowledge level of individuals, which includes the instinct to save themselves (Leykin et al., 2013). Second is the preparedness for a disaster (Gunderson, 1999; Smith et al., 2012; Sterling, 2010). Third is the ability to learn and innovate (Keim, 2008; Newman & Dale, 2005; O'Rourke, 2007; Paton & Johnston, 2017). Fourth is the flexibility, such as rearranging commutes according to the recommendations of the meteorological department (Guiver, 2013; Rogers & Tsirkunov, 2013).

2.3 Relationship Among Individual Resilience, Human Behavior and Potential Determinants of Individual Resilience

Based on the foregoing, individual resilience is an adaptive cycle and loop that can be changed through learning and nurturing. In order to learn to nurture resilience of people, it is necessary to identify the relationship among individual resilience and human behavior, which can help to find determinants of resilience.

2.3.1 Human–Environment Interactions

Assumptions of human–environment interactions are well suited for individual resilience research. Human–environment interactions integrate the dynamic and complex interplay among environmental, ecological, and sociopolitical components of human–environment systems. It involves multiple temporal and spatial levels, which incorporate individual resilience quite well (Cutter et al., 2008; Frazier et al., 2013; Hughes et al., 2005; Stokols et al., 2013).

Social ecology is the theoretical foundation for human–environment interactions (Pálsson, 1996; Penn & Mysterud, 2007). Social ecology centers on the processes involving transactions among assets, resources and actors (Stokols et al., 2013). Sociologists at the University of Chicago during 1920s and 1930s, began to study the way in which the spatial distribution of land use and resources may influence behavioral disorders (Bennett, 1996; Stokols et al., 2013). Physical and social relationships are then developed and human ecology theory emerges, integrating physical development with the urban planning, sociology, geography, anthropology and psychology (Altman & Werner, 2013; Dunlap et al., 1979; Slocombe, 1993). Then social ecology is provided based on the human ecology theory and works as the foundation to study on the human–environment interactions (McGinnis & Ostrom, 2014; Stokols et al., 2013). Since then, human beings and their behavior are increasingly focused on; particularly, how the surroundings influence and make human beings behave differently need a concern (Altman & Werner, 2013; Van Wormer, 2017). Finally, scholars suppose and further prove that environment, including the physical environment and social environment can influence human behavior (Hess et al., 2012). More specifically, human behavior and human behavioral adjustment—individual or collectively—are cognitions and perceptions of the environment (Brown & Westaway, 2011). That is the assumptions and content of theory of human–environment

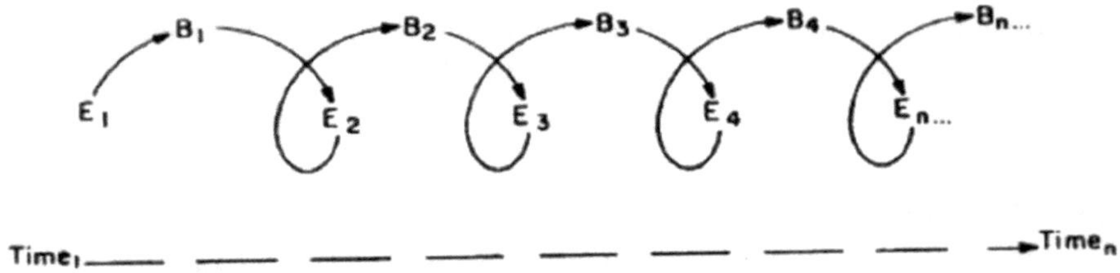

(a) Resilient Environment-Behavior Systems

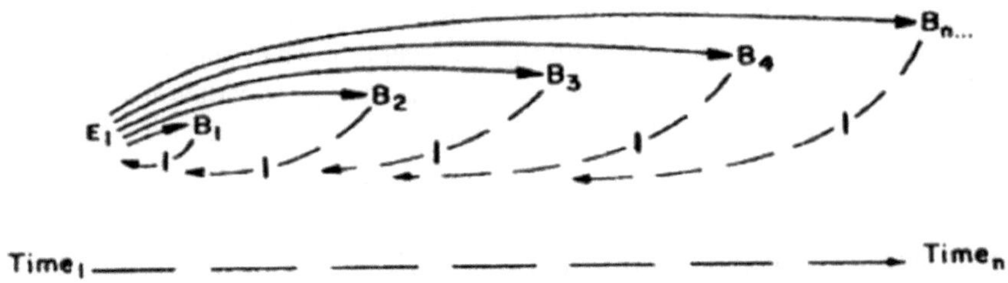

(b) Non-Resilient Environment-Behavior Systems

Fig. 2.3 Schematic representations of human–environment interactions. *Source* Cited from Stokols et al. (1978, 2013)

interactions. Figure 2.3 shows the interactions among different groups of individuals living in different kinds of environment.

Figure 2.3 shows different interactions between human and environment, reflected in different human behavior in what are termed environment-behavior (E-B) interactions. Environmental changes can elicit a variety of individual, organizational or societal responses. For a resilient system (Fig. 2.3a), individuals or groups may improve their behavior in their socio-physical environment associated with climate change, with the aim to improve the level of congruence or fit between themselves and their surroundings (Stokols, 1978; Stokols et al., 2013). The root cause for this type of behavior change is individual or collective subjective understandings and perceptions of the situation. Therefore, it is a behavior pattern that it can evolve and adapt.

In less resilient systems (Fig. 2.3b), human beings cannot modify their surroundings in an adaptive pattern because the physical and institutional environments are so rigid. Under these conditions, human behavior can change (at times $1-n$); but not in an adaptive fashion, and the personal or collective behavior remain captive to the dictates of rigid environmental conditions (E1). Therefore, there are no efficient and adaptive responses of individuals or groups between environmental changes and changes in human behavior (Stokols et al., 2013).

2.3.2 Human–Environment Interactions and Potential Determinants of Individual Resilience

Individual attributes The existing literature verifies and identifies attributes of individuals that influence and determine resilience, such as gender, age, faith, income, leadership, self-efficacy determine residents' level of social resilience (Cohen et al., 2016; Cutter et al., 2008; Gaillard, 2007; Gunderson, 2010; Leykin et al., 2013; Nguyen & James, 2013; Soetanto et al., 2016).

Living environment Living environment is such an important determinant for individual resilience. First of all, the physical environment including the built and natural environments can afford an individual living place's functioning and survival during an external disturbance (Ahern, 2011; Smith, 2014). Then, social environment including the management modes adopted by the formal and informal intuitions are also closely related to individual resilience (Hess et al., 2012; Paton et al., 2000). During an external disturbance, they work as the brain and direct the individuals to behave, e.g. responding to and learning from disturbances.

Perception of disaster risk Disaster risk perceptions of individuals are proved to be factors influencing human behavior towards disturbances (Burton, 1993; Izadkhah & Hosseini, 2005). Previous experiences closely related to disaster risk perceptions are also highlighted for it influences human behavior towards disturbances (Henly-Shepard et al., 2015; Paton et al., 2000; Shaw et al., 2004).

Perception of living environment Based on the theory of human–environment interactions, we can conclude that the interactions between individual and environment can influence human behavior, including resilience-oriented behavior; therefore, these interactions can influence individual resilience. Existing studies can prove the interactions between individual and environment can influence human behaviors towards natural disturbances. Sense of place is claimed and argued to be a key factor in disaster recovery (Chamlee-Wright & Storr, 2009; Silver & Grek-Martin, 2015). Recently, psychological attributes of individuals are also focused on for their influences to human behavior towards disturbances (Brown and Westaway, 2011; Hess et al., 2012; Paton et al., 2000). All these research support that attributes of personal perceptions should influence human behavior towards natural disturbances. However, existing research have not linked human behaviors, resilience and this type of determinants together. In other words, the subjective factors of individuals to their living environment as determinants of individual resilience have not been studied widely and deeply.

In summary, this section of literature reviews the conceptual relationships between individual resilience and personal perceptions of their living environment. Human–environment interaction connects them together. It turned out that most resilience research only focused on the objective attributes about the individuals as the potential determinants of individual resilience. However, the subjective factors, especially the ones reflecting the personal perceptions of their living environment have not attained enough attention, which are probably determinants of individual resilience based on the theory of human–environment interactions.

2.4 Relationship Between Resilience Nurturing and Urban Management

Based on the above, resilience is a changeable process, which can be reflected through human behavior. Through the human–environment interactions and environment-behavior of individuals, we can conclude that efficient external interventions (e.g., policies) may make human behavior more resilient if the institutional environment is not constrained (Glendon et al., 2016; Michie et al., 2009; Peterson, 2000). Moreover, through changes in the physical, social and institutional environment, personal perceptions and understandings to them can change. So a positive change in the environment probably contributes to the enhancement of people's resilience (Stokols et al., 2013).

In the fourth phase of the adaptive cycle, namely the reorganization stage, learning is the pathway to the realization of adaptation. This suggests that learning may change human behavior towards more resilient ones. Regarding it, the concerns and behavior of governments should be highlighted as a powerful, positive intervention to nurture individual resilience. This means that when the behavior of individuals within a community is counter to those required for resilience to urban flooding, they are constrained by the political processes, e.g. urban management.

2.4.1 Urban Management for Resilience Nurturing

Urban management is defined as the responsibility of municipal government with the aim to work for the attainment of sustainable urban development addressing the ecological, social, economic, cultural, environmental and institutional sectors (Amos, 1989; Wong et al., 2006; Zamani & Arefi, 2013). It is widely accepted as functions of government, including policy formation, social services, resource allocation, implementation and operation (Cheema, 1993; McPhearson et al., 2014; Rakodi, 1991). McGill (1998) points out urban management does not only have strategic (to plan) but also has the operational (to provide and maintain) functions.

Debates on management on resilience nurturing to natural disturbances have flourished since the 1970s (Gaillard, 2010). Discussion suggests that low resilience or called low adaptation capacity to natural disturbances of individuals should be attributed to poor urban management (Fekete et al., 2014; Quarantelli, 1988; Turner, 1994). For example, economic pursuit rather than environmental considerations, government corruption, and over-regulation by rigid bureaucracies cause problems for any type of urban management, including the resilience nurturing (Ahrens & Rudolph, 2006). The fragmentation of disaster-related sectors also increases challenges to traditional disaster management, namely prediction and control management, which focuses more on technical solutions without considering the interactions among environmental, social, and institutional dimensions (Jiang et al., 2013; Pahl-Wostl, 2007). In contrast, successful urban management mostly helps failed socio-ecological systems to enhance, build, or develop resilience capacity (Buckle, 2006; Masten, 2015).

2.4.2 *"Learning-By-Doing" Government for Resilience Nurturing*

A "learning-by-doing" government is that nurtures stakeholders, especially those living in vulnerable areas, to attain the desired resilience under the urban management (Walters & Holling, 1990).

Social learning theory links urban management to resilience nurturing. Berkes (2007) noted that specific linkages based on social learning consist of (1) learning to live with change and uncertainty, (2) nurturing various types of ecological, social, and political diversity for increasing options and reducing risks, (3) increasing the range of knowledge for learning and problem-solving, and (4) creating opportunities for self-organization, including strengthening of local institutions and building cross-scale linkages and problem-solving networks.

Learning theory has long been understood as an integral part of resilience thinking (Tschakert & Dietrich, 2010). Folke et al. (2003) stated that learning based on diverse social memory in tune with ecosystem dynamics could build resilience. Davidson-Hunt and Berkes (2003) also mentioned that the individual and institutional levels of learning and their dynamic interactions with ecosystems are the foundations for coping with and adapting to changes. Based on the literature, learning is a key aspect of resilience theory. It connects the realization of adaptation to disturbances and stakeholder participation, which is closely related to nurturing resilience and decision-making processes (Newig & Fritsch, 2009; Reed, 2008).

According to learning entities, including individuals, groups, organizations (informal ones), and governments (formal ones), pathways for learning can be classified as self-learning (individuals), collective learning (mutual learning within groups) (David et al., 2016), and government learning (Etheredge & Short, 1983). Social organizations learn and help individuals through social or collective learning. Government institutions learn and help organizations and individuals learn through institutional or government learning.

Learning as human behavior includes the active and passive learning facilitated by different motivations. Personal qualities of the facilitator, e.g. individual disaster awareness, trust, and willingness to establish alliances are important for self-learning (Mostert et al., 2007). However, once facilitators, such as disaster awareness for the individuals are missing, the facilitators should be well nurtured in group interactions and have skills and expertise in collective learning. Motivations for collective learning include defending interests and providing social services (Mostert et al., 2007). However, once collective facilitators, such as social networks, are missing, the facilitators should be well-nurtured by welfare governments and motivated by governmental responsibility and defending their political achievements and interests (Crijns et al., 2017).

Learning in resilience refers to social and governmental learning, which can be referred to as "learning-by-doing" (Berkes, 2007). Higher levels of learning are important, as they can compensate for deficiencies in self-learning or collective learning (Olsson et al., 2004). Governmental learning is important for the efficient intervention and decision-making processes that shape adaptation and resilient livelihoods beyond climate disturbances. It serves as the set of social memory, which is based on previous experiences (Adger, 2003). The previous experiences of institutions help reorganize conceptual models and paradigms based on the constant enhancement of understanding of the methods to cope with and adapt to disturbances. Therefore, except for learning from the past, forward-looking (anticipatory) learning is needed for institutions for their adaptation and resilience in the context of climate change (Tschakert & Dietrich, 2010), such as the new, advanced, modern concepts and technologies institutions learn to control.

Most scholars frame social learning alongside governments with learning loops (cycles). The literature describes three learning loops (cycles) associated with government learning (David et al., 2016; Tschakert & Dietrich, 2010). *Single-loop learning* refers to improving instruments for better management, such as improving infrastructure for better urban flooding management (Maarleveld & Dabgbégnon, 1999). *Double-loop learning* refers to radical behavior changes based on changes to underlying values and beliefs, such as focusing on ecological environment protection without pursuing only rapid urban development (David et al., 2016; Pahl-Wostl, 2007). *Triple-loop learning* is concerned with governance protocols that can protect the realization of single-loop and double-loop learning. It involves learning processes in the political context and the notion of power (David et al., 2016; Pelling et al., 2015). Cases suggest institutions emerge to help cope with disturbances, and these are then reshaped by the disturbances (Folke et al., 2005). Government learning "may be bridged by allowing for and encouraging reflective practices that can lead to transformative learning, which is defined as leading to a change in an individuals' frame of reference, with potential consequences on the individual's behavior" (Cranton, 1994, 1996; Mezirow, 1991, 2000; Pelling et al., 2015). Transformative learning can bring about behavior changes in individuals, because it help them reshape and delimit individual feelings and perceptions, and cognition by predisposing intentions, purposes, and expectations (Gunderson, 2010; Mezirow, 2006; Pelling et al., 2015; Sharpe, 2016).

To conclude, social learning is a necessary path for putting resilience into practice. It occurs at different scales and comprises three loops. Triple-loop learning protects the single and double loops. Social learning, especially in governments, facilitates the desired outcomes of social learning processes as individuals, groups, organizations, and even governments move beyond transferring knowledge to inputting it. Therefore, positive of this type of learning can make individuals more resilient to natural disturbances.

2.4.3 *Place-Responsive Urban Management*

Substantial efforts on management have been made by to reduce the impact of hazards (O'Brien et al., 2006). It has been proved that the strategies for DRR should be place-responsive (Cash et al., 2003; Cutter, 2012; Cutter & Finch, 2008). Researchers are studying place-responsive emergency management for vulnerable areas to natural disturbances. Cutter and Finch's (2008) study of areas vulnerable to natural hazards suggested community-based policies and emergency plans. Cutter's (2012) detailed studies on environmental justice considered that policies for disaster management on a large scale might exacerbate inequality. Cash et al. (2003) remarked that the place-based knowledge should be translated into the scientific plans and policies for better risk reduction. We can find that most of place-based management research focuses on emergency management; however, the place-based urban management in terms of human–environment interactions has been neglected. Therefore, this research wants to fill the gap.

To summarize, social learning is a path for nurturing individual resilience. Through successful social learning, especially in governments, the desired outcomes of social learning processes and the various actors can move beyond transferring knowledge to inputting it. That is why the urban management can nurture individual resilience. Besides, the place-responsive urban management is proved to be efficient for DRR, which is the theoretical supporting for this research on finding place-responsive strategies of urban management to improve individual resilience performance.

2.5 Conceptual Framework

Above-mentioned section summarized the correlations among resilience, individual resilience, determinants of individual resilience, and resilience nurturing under urban management through social learning. The concepts and definitions of resilience, including adaptive cycle theory, resilience capacity, social resilience and individual resilience were explored. The relationship between the resilience performance and human behavior, embedded into coping and adaptive strategies, is reviewed. Then the determinants of individual resilience in terms of personal perceptions of their living environment are conceptually reviewed in terms of human–environment transaction theory. Furthermore, social learning theory is reviewed to support that political intervention from the urban management can build resilience, however it requires the intervention should be on a place-responsive basis.

2.5.1 Research Gap

Human resilience-oriented behavior is focused on in this research, which can be identified and judged to evaluate individual resilience performance. This method has been widely accepted and applied into the human health research,[2] however have not seen on the research of resilience to natural disturbances. This research creatively applied this method to evaluate individual resilience based on the adaptive cycle theory, to fill the gap on quantitative assessment of resilience.

In-depth, longitudinal research is conducted at the scale of individual resilience to reflect community resilience. Individual indigenous and monitoring observations are important because they highlight the subjective factors for resilience performance, which are seldom integrated into current frameworks and models. This research addresses the gap in global science and provides insights regarding the situation in China.

This research examines the effect of subjective factors reflecting personal perceptions of living environment to resilience performance. Socio-demographic attributes of individuals and living environment of individuals are the main focus as the determinants of individual resilience. However, scholars ignore the individual perceptions of their living environment (both social and physical environment), which are also potential determinants of individual resilience performance.

Finally, place-responsive strategies for local social groups to nurture resilience have been neglected. The interventions are always made on the city or national scales, at the expense of place-responsive management for individual resilience nurturing, especially for the low-income residents and communities. This research is based on the examination and exploration of the individual resilience and its determinants, with the hope of nurturing individual resilience for the special group of population, which will enrich the theoretical knowledge of these processes in China.

2.5.2 Conceptual Framework

Based on the conceptual review, in order to fill these research gaps, this study is framed as follows (Fig. 2.4).

The research frames resilience as the correlation between individual resilience and urban management. In essence, individual resilience is reflected by the coping and adaptive capacity of individuals. As a result, people's perceptions, including perceptions to natural disturbances, the physical environment and social environment are identified and evaluated as the potential determinants of individual resilience. Meanwhile, urban management as an external intervention can change human behavior to be more resilience-oriented. More specifically, based on human–environment interactions, this research argues that urban management plays a positive role in enhancing the physical environment and social environment, facilitating the changes of people's

[2] http://www.measuringimpact.org/s4-sf-36.

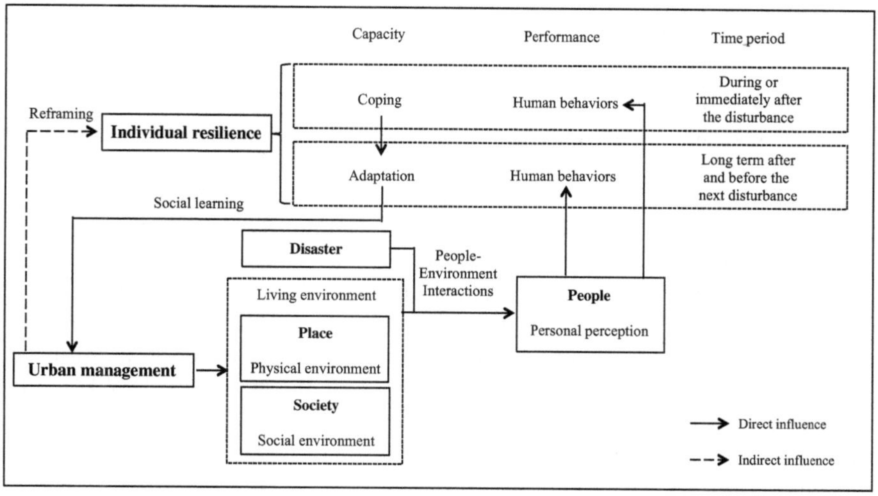

Fig. 2.4 Conceptual framework. *Source* Author

perceptions to the living environment and even to natural disturbances. Eventually this positively affects their resilience performance.

The evaluation of resilience performance and its influencing factors are conducted separately. For resilience capacity, the performance of human behavior, embodied in the adopted strategies of coping and adapting to natural disturbances, is identified and judged. Several people-centered subjective factors reflecting personal perceptions of natural disturbances, the physical environment and social environment (People-Disaster-Place-Society interactions) are potentially critical determinants of resilience performance. The correlation between resilience performance and these personal perceived factors can then be explored on a place-specific basis, identifying the significant determinants of individual resilience performance of selected study area.

Since individual resilience can be nurtured towards the positive direction, urban management strategies are necessary to nurture resilience for reducing the potential risk from natural disturbances. Social learning and its loop theory can be adopted to explain the changes of individual resilience associated with the external intervention. The government, as a powerful actor in possible interventions, can improve individual resilience performance within the human–environment interactions. As such, social learning connects resilience improvement and urban management. Based on these, place-responsive policies and measures for nurturing individual resilience can be provided based on the social learning of the people-disaster-place-society interactions. More specifically, urban management for individual resilience improvement can be implemented through the improvement of communities, including the physical environment and management, which are closely related to the perceptions of individuals. This provides another pathway to achieve improvements in resilience.

2.6 Conclusion

This chapter reviews the recent theoretical and applied research on resilience and urban management, as well as the relationships among them. The way in which the social learning theory connects these concepts is reviewed based on people-disaster-place-society interactions. Finally, to understand the concept of individual resilience, a conceptual framework is built to examine the determinants of personal perceptions of natural disturbances, physical environment, and social environment.

In addition, the definitions and related theories of resilience are summarized, and the concept of resilience in this research is narrowed down to individual resilience. Then the relationship between human behavior and resilience performance is reviewed from the livelihood perspective, forming the theoretical foundation for the assessment of resilience performance. Implicit in the first research hypothesis is that there are significant differences in individual resilience to urban flooding for residents in different communities.

The people-centered subjective factors reflecting personal perceptions of natural disturbances, physical environment and social environment are summarized to identify their relationship with the current state of individual resilience. Then the second hypothesis of this research is that individual perceptions of urban flooding, living place and living society can influence individual resilience.

Finally, the conceptual relationship between the government interventions and human behavior is identified from literature, suggesting that government efforts can change individual behavior. Place-responsive urban management and individual resilience are linked by the social learning theory through human-environmental interactions. This forms the basis of the third research hypothesis that individual resilience cannot be nurtured through resilience-oriented measure, i.e. Sponge City Initiate from central government of China.

Chapter 3
Empirical Background: The Causes, Threats of, and Responses to Urban Flooding in China and the World

3.1 Introduction

Climate change is a global challenge, and cities are experiencing extreme weather events, such as heavy rainfall more frequently (Arnell & Gosling, 2016; Cannaby et al., 2015). Climate-driven disasters are increasingly serious threat associated with climate change. Among the climate-driven disasters, floods are the most frequent and widespread ones associated with extreme weather, e.g. heavy rain (Dhar & Nandargi, 2003). Urban floods, which may be several feet deep, occur quickly after intensive and heavy precipitation deserves China's attention because of the frequency and severity of its urban floods.

In this chapter, the causes, impacts and measures of urban management on urban flooding mitigation in China and the world have been reviewed. Following them, the focused city Shenzhen, China, is carefully introduced on its current situations of urban flooding, impacts from urban flooding, as well as the measures of urban management for dealing with such problems.

3.2 Causes of Urban Flooding

3.2.1 Urbanization-Led Climate Change: Extreme Rainfall and Urban Flooding

Urbanization transforms original natural land into land suitable for human habitation, and the effects of microclimates and thermal islands associated with urbanization exacerbate urban climate change, and the frequency and intensity of precipitation events (Andjelkovic, 2001; Ye et al., 2016). The relationships and analyses of urbanization, climate change, and urban flooding have been reported in numerous

J. Song, *Individual Resilience to Urban Flooding and the Implications for Urban Management*, https://doi.org/10.1007/978-981-97-0546-7_3

theoretical and empirical studies (Asif et al., 2015; Srinivasan et al., 2013; Wang et al., 2014).

Rapid urbanization has been proven to be associated with climate change, in terms of changes in hydrological and hydro-meteorological processes, and it can influence the frequency and intensity of storms (Jung et al., 2011). Urbanization damages natural land cover, discharges anthropogenic waste heat and suspended particles, and precipitates microclimate changes in urban areas (Wang et al., 2014). It leads to temperature, hydro-climatological and hydrological changes, and disrupts the balance of water ecosystems, precipitates intensive and frequent rainfall events, and increases the risk of urban flooding (Wang et al., 2014). Urbanization also creates urban heat islands (UHI) and cities' underlying surface conditions influence physical and atmospheric urban environments' convective weather patterns and changes in precipitation (Lin & Yu, 2005; Liu et al., 2007). UHI directly influence climates in urban areas, creating microclimates that lead to heat-island-initiated precipitation (Dixon & Mote, 2003). In summary, water ecosystems have been transformed as a result of climate change and associated microclimatic changes (Huong & Pathirana, 2013).

Numerous empirical studies have also verified the relationship among urbanization, rainfall extremes, and urban floods. Urban floods are frequent, since heavy rainfall is associated with rapid urbanization (IPCC, 2007). Harbor (1994) has noted that land-use change will increase the extent of impervious areas to rainwater, which will result in increased surface water runoff. In India, Shastri et al. (2015) concluded that rapid urbanization has led to regional rainfall patterns and extreme weather, based on a detailed case study of Mumbai. Zheng et al. (2013) also used precipitation data for Beijing from 1971 to 2010 to verify the relationship between the city's extreme rains and urbanization; results show that urbanization leads to heavier rainfall in downwind areas. Yang et al. (2013) analyzed the relationship between urbanization and the frequency of urban flooding occurs, and found that spatial and temporal changes associated with urban flooding are related to rapid urbanization. Shi et al. (2007) used Shenzhen as a case and concluded that land-use changes have played a dominant role in changing surface runoff over the past 30 years. In southeast Texas, Burian and Shepherd (2005) found that changes in precipitation were most significant in urban areas, especially in afternoons during the warm season. Guo et al. (2014) used models and empirical cases to verify the discharge of aerosols influences precipitation differently, depending on the amount of rainfall. They proved the average precipitation after pollution would increase by up to 17% on cloudy days.

Empirical models have verified some theories. These models simulate relationships between climate change and urban flooding, based on land-use changes that are closely related to urbanization. Some models are used to simulate urban rainfall extremes change caused by anthropogenic-induced climate change (Arnbjerg-Nielsen et al., 2013). Alternatively, some are used to simulate future climate change scenarios by predicting different CO_2 concentrations (Feyen et al., 2009), or by simulating extreme and intensive precipitation events that increase the risk of urban flooding (Veronesi et al., 2014). Rainfall models are also used to simulate the relationships between climate change and urban flooding. Land-use simulation models have found that urban growth was projected on the basis of historical growth patterns (Dinamica EGO). Particularly, WRF is applied to explore the relationship between rainfall patterns to urban heat island effects (Noah LSM). SWMM was applied to simulate surface inundations and storm-sewer surcharges, to quantify the increase in flood hazard that would result from the changes. The changes are that climate change, combined with urban growth, has significant implications for urban flooding (Huong & Pathirana, 2013).

3.2.2 Urbanization-Led Construction: Human Activities and Urban Flooding

Urban areas are more likely than the countryside to experience floods, for land-use changes and the human activities associated with urbanization are known to play a dominant role in precipitating urban flooding, from the perspective of flood discharges and the increased frequency of flooding (Konrad, 2003; White & Greer, 2006).

Today, more people are migrating from rural to urban areas. More than half of the world's population lives in cities (UN, 2006), and disasters affect concentrated populations more severely. There is no doubt that concentrated city populations are more vulnerable to disasters, including floods (de Loyola Hummell et al., 2016; Nicholls & de la Vega-Leinert, 2008). Meanwhile, there has been a sharp increase in human settlements, infrastructure development and industrial growth associated with urban expansion (Huong & Pathirana, 2013; UN, 2006). Roads and buildings replace vegetation and soil. The replacement process increases the likelihood that there will be more flooding of lands, which collect, absorb, and store rainwater—such as forests, grasslands, and water systems. Therefore, the increasing extent of man-made surfaces leads to more urban flooding (Mote et al., 2007; Shepherd et al., 2002). In his master's degree book, Koster (2013) noted that urbanization has led to an increase in surface runoff in the Playa catchment, Bonaire. In addition, soil erosion caused by human activities will lead to urban floods. Soil erosion will reduce water's absorption, and increase the runoff to ditches and streams. It will also change the soil, which will clog drainage systems and inlets (Andjelkovic, 2001). Bhatt et al. (2014) found that high rates of soil erosion increased the volume of upstream dissolved substances and sediments in river loads deposited in channels, reducing the drainage capacity of natural systems. Some construction activities alter a city's land surface, as the lower areas are more readily flooded (Hollis, 1975; Sakieh et al., 2015). To conclude, urban construction activities increase the likelihood of urban flooding, since the ground areas in these places are usually pitted.

Natural water systems are destroyed in the course of urbanization. The construction of cities has destroyed the original rivers, lakes, wetlands, ponds, and channels. Bodies of waters have been reduced, as have the connectivity of water systems and the available rainwater storage capacity. Ji et al. (2014) used remote sensing images to find that over the past 30 years, there has been a 41.03% decrease in river length, and the number of rivers in the study area has decreased. Han et al. (2016) pointed out that the Yangtze River Delta network had been degraded by urbanization. The stability and complexity of streams have changed with urbanization. Han et al. (2013) verified that urbanization reduces the connectivity, composition, and structure of river systems, creating conditions that will lead to urban flooding. Peipoch et al. (2015) showed that human activities impact river networks, which in turn will increase the possibility of urban flooding by reducing and disrupting the rivers' ecological functions.

In addition to destroying the natural landscape, of course, human beings construct the infrastructure needed to solve rainwater transportation problems. People have always fought floods with infrastructural components such as levees, dams, reservoirs, and modern sewage systems. Today, sewage systems are the main type of infrastructure used to curtail urban floods. However, not even the most advanced sewage systems prevent them. China's sewage systems have many problems. The design of and planning for drainage systems is not as mature as in developed countries, which usually have multilevel drainage systems (Kang, 2014; Li et al., 2015). Sewer pipe design standards are usually not achieved. Liu et al. (2014) wrote that the drainage pipes in most of China's cities cannot meet the design standard of 1-in-1 year rain frequency, and 90% of old towns cannot even meet the minimum design standards. Hao et al. (2014) observed that the pipes in the old towns of China are also old, and most do not meet society's modern drainage code requirements. Furthermore, the management of sewage systems is inadequate (Kang, 2014). In summary, China's sewage problems are caused by its low capacity, because of poorly connected sewage pipes. These are longstanding problems that are difficult to solve.

Apart from climatic events, rapid urbanization is the main cause of urban flooding. However, we cannot control the speed of urbanization. Therefore, the challenge of balancing the relationship between urban development and urban flooding will not go away. As China is the largest developing country in the world, studying its urbanization will illuminate the relationship between development and urban safety. For this reason, this study seeks a theoretical and practical pathway for solving the problems of urban flooding.

3.3 Threats of Urban Flooding

In the wake of urbanization, populations and resources are highly concentrated in cities. Changing demographics characterized by ageing populations are increasing the vulnerability to urban floods with their environmental, economic, and social impacts.

3.3.1 Impacts on Habitat

Serious environmental effect of urban flooding has been an alarming for the increase of water pollution (Izzati, 2016; Owa, 2014; Sartor et al., 1974). Storm water may contain bacteria, organic waste, suspended solids, nutrients, animal waste, heavy metals and even sediment from construction sites (Mason et al., 1999; Owa, 2014). When wastewater spills into city streets, they not only threaten the living environment of wildlife, but may also pollute local waterways and even drinking water. Urban floodwaters mixed with household, industrial, and agricultural garbage pollute the environment and threaten human health (Andjelkovic, 2001; Xie, 2013). Floods also negatively influence the surrounding ecosystems, and given that the city itself is a vulnerable socio-ecological system, prolonged flooding is dangerous to all living things.

3.3.2 Impacts on Economy

Urban flooding results in huge economic losses. On July 21, 2012, urban floods in Beijing damaged 10,660 properties, and caused CNY 11.64 billion in economic losses.[1] In October 2013, Typhoon Fitow struck Ningbo. The city suffered 5720 damaged properties, and CNY 2 billion in economic losses.[2] In the past five years, economic losses caused by urban flooding in Guangzhou each year have exceeded CNY 500 million.[3] Flooded streets cause traffic delays that will result in further economic losses for industries that depend on access to transportation.

3.3.3 Impacts on Society

A city can recover economically from a disaster, but it is hard to repair the damage to a society. In 2012, urban floods in Beijing killed 79 people. Many scholars have pointed out that urban flooding will harms people's mental health (Ahern et al., 2005), and will at the very least limit their ability to study or work. Fewtrell and Kay (2008) presented evidence that constant urban flooding increases the likelihood of physical and mental health problems. In addition, urban flooding leads to the dangerous leakage of electricity to threaten human health (Zhu, 2013). Zhu (2013) observed that the floating sewage covers for rainwater intrusion into the sewerage systems are a potential hazard to pedestrians. To conclude, the social impact of urban floods includes delays in transportation and the disruption of water, sewage and electrical services.

---

[1] goo.gl/1auswW.

[2] goo.gl/KJuNuJ.

[3] http://shizheng.xilu.com/20140105/1000150000508397.html.

3.4 Responses to Urban Flooding: China and the World

It is necessary to design strategies that would be helpful in overcoming or reducing losses to urban flooding. Governments, scholars and engineers have tried to reduce flood hazards. From a vertical perspective, strategies for overcoming urban flooding include problem identification, infrastructure enhancements, and disaster management.

3.4.1 Developed Countries

Developed countries are leading the fight against urban floods (Baker, 1977; Cembrano et al., 2004; Hey, 2002; Nie et al., 2009). Some of these strategies can be emulated by developing countries.

Developed countries improve their drainage systems. In the United States, Chicago was quick to recognize the importance of urban flood management, and the municipal government spent more than USD 50,0000 a year to clean and improve the city's drainage systems. The government also developed the Tunneland Reservoir Plan (TARP) in the 1970s—to be completed in 2029—to prevent rainwater and sewage in the sewage system from flowing into the urban river system. In 2007, London spent USD 1.7 billion to enforce the Thames Tunnel plan, the construction of a 35-km-long drainage tunnel with a diameter of 7.2–9 m. This tunnel will improve the city's drainage capacity, and make the city safer during storms. Australia separates its drainage systems into big and small systems (Booth, 1991). Small drainage systems include road drains or pipes and networked systems, and are used to manage small volumes of storm water runoff. Big systems include surface drainage or runoff channel storage systems, and are used to store rainwater after its volume has exceeded the storage capacities of the small systems (Nickel et al., 2014).

Some developed countries are using ecological methods to improve their drainage, using technologies such as porous pavement road surfaces and some green and blue lands, such as forests, wetlands, and rivers, which not only help to drain rainwater but can also be used to store rainwater (Booth, 1991; Nickel et al., 2014). This idea is called *low impact development* (LID). It has been adopted by many countries and cities (Ellis, 2013; Montalto et al., 2007; Zahmatkesh et al., 2014). Germany's greening rate is quite high, and green lands can absorb rainwater and reduce stormwater runoff. Holland converts its surplus land to wetlands and polders, low-lying lands to lakes and detention basins, strengthens temporary channels, and widens and deepens riverbeds to improve the rivers' storage and drainage capacity. Holland has changed its original drainage mode from a single mode to multiple modes: retention, storage, and disposal. As a result, wetlands, polders, low-lying lands, lakes, and channels will be used. When there are storms and floods, these places absorb the water. As these sponges are inter-connected, they will drain off the water into an outside water system, and reduce the possibility of urban flooding.

There are also some comprehensive methods for urban flood prevention. For example, Japan uses its parks and sporting grounds as temporary ponds for storing rainwater (Suzuki et al., 2002).

Except for the physical and ecological infrastructural improvements aimed at fighting urban flooding in developed countries, urban management for reducing the impacts from floods is focused on. Evaluating the impacts of rainwater and the hazards associated with urban floods is the foundation of urban flood management. Australia, Japan, the European Union, and the United States, have all established management mechanisms for evaluating the impacts of rainwater and urban floods (Burby, 2001; Ikeda et al., 2008; Wheater, 2006). Accurate and timely assessments will monitor the distribution of urban floods, simulate and model the possibilities of occurrence, the periods of submergence, their range and depth, which will enable an assessment of whether urban areas are in danger (Cattoën et al., 2016; Yucel et al., 2015). At the same time, it will be possible to judge whether construction activities are likely to create hazards for residents a storm hits (Alfieri et al., 2013; Terti et al., 2015).

Financing mechanisms are another important aspect of urban flood management. Since the construction of an urban flood management system is a public endeavor, governments are the main sources of funding. It is therefore necessary to establish legal accountability. By law, governments need to allocate some percentage of their revenues to improving and maintaining the infrastructure needed to reduce urban floods. As a result, the implementation of urban planning is protected, and industries are motivated to participate in the construction of infrastructure, since they will benefit from its construction. Once enacted, the law must be enforced. For example, flood insurance is promoted by the United States for the management and planning of floodplains (Burby, 2001).

In developed countries, containing urban floods reductions is comparable to project training material for disaster reductions, especially with regard to the elderly and people with special needs, who are at greatest risk during disasters (Benson et al., 2007; Sullivan & Häkkinen, 2006; Tran & Shaw, 2007). Japan is famous for its rigid and high-quality training and disaster drills (Bajek et al., 2008; Norio et al., 2011). NGOs and other non-state industries participate in the training and in other disaster mitigation actions (Benson et al., 2007; Twigg & Steiner, 2002).

In developed countries, measures for treating the mental and psychological problems that follow disasters have received attention. Fewtrell and Kay (2008) used a case in the UK to quantify the health impacts of flooding, and called for attention to be given to the mental health of flood survivors. Hajat et al. (2005) expressed the need to emphasize mental problems after urban flooding, and observed that aid should shift from post-disaster improvisation to pre-disaster planning.

In summary, different governance models have informed measures for mitigating the impacts of urban flooding. Although the objectives are the same, the forms of the organizations, policy-making processes, legislation, and public participation vary from area to area. Developed countries have implemented rounds of management enhancement to combat urban floods: improved drainage systems, ecological infrastructure construction, evaluations of the impacts of urban flood system construction, legislation, and policy changes (Tunstall et al., 2004). Other comprehensive methods, such as the participation of NGOs, insurance encourage, have been also been adopted to reduce the hazards associated with urban floods.

3.4.2 Developing Countries

Tingsanchali (2012) used the example of Thailand to verify the effectiveness of a conceptual framework on integrated flood management that includes four cyclic steps: (1) forecasting and an early warning system; (2) preparing; (3) developing emergency responses; and (4) effecting recoveries and rehabilitation. Montoya (2003) suggested updating the building and lifeline geo-databases for urban flood management, and tested it in the Costa Rican city of Cartago. Tingsanchali (2012) observed that in most developing countries, the government is responsible for flood disaster management; the participation of nongovernmental agencies and private sectors is limited. He concluded that urban flood management implementation activities are not coordinated or integrated (Tingsanchali, 2012). Ward et al. have stated that governance plays the dominant role in flood management, and developing countries' governance lessons must take place at the municipal level. Veronesi et al. (2014) made the important observation that residents of developing countries lack hazard awareness, and do not want to pay to reduce the ecological and health risks of flooding. Gupta (2007) combined resilience theory with urban planning and management, and formulated some lessons for future urban flood management in Mumbai.

Several measures have been updated and improved to solve the problems facing developing countries as a result of more frequent and heavier rainfall. They are learning about the advanced measures used for urban flooding mitigation in developed countries, but the countries' different governance and management strategies are of limited value in developing countries. Therefore apart from making full use of modern technologies, developing countries should combine the advanced concepts used in developed countries with their own realities, and seek improvements from the perspective of developing better governance arrangements for managing urban floods.

3.4.3 China

In China, many cities suffer urban floods every year, which are costly and sometimes deadly. Governments and scholars in China and around the world are trying to find solutions. Sewage systems in China exist many problems like old with low drainage capacity. Poor man-made sewerage system with increasingly loss of natural sewerage system is one of reasons there are urban floods.

Many scholars have offered ecological wisdom to solve Mainland China's problems. *Ecological wisdom* is "a peculiar form of ecological domain knowledge that comprises specialized instances of one individual's or one group's prior and time-honored knowledge in the realm of ecological research, planning, design, and management (Xiang, 2014, p. 67)." The Sponge City Initiative of China is based on ecological wisdom, and LID concept combined with ecological wisdom is applied to create a sponge city.

At the National Urbanization Conference on December 12, 2013, President Xi expressed his intention to build sponge cities that can store, drain, recycle, and purify rainwater using the natural infrastructure.[4] On October 22, 2014, the Ministry of Housing and Urban–Rural Development of the People's Republic of China (MOHURD) published the Guidelines for the Sponge City Initiative and Guidelines for realization of Sponge City Initiative (State Council, 2015).[5]

The guideline defines a sponge city as a city that can withstand climate change and natural disasters. It can absorb, store, drain, and purify rainwater, then release and recycle the rainwater as needed. Several zones of a city have been selected for the sponge bodies (LID practices) comprised of sponge city, which will be built from permeable materials or green lands that can absorb precipitation. When there is no rain, these lands can be used as parks or gardens, or as some other public green space.

There are three types of sponge cities. Yu compares a sponge city to a sponge that can control flooding (Chen & Yu, 2015; Yu et al., 2015). Qiu (2015a, 2015b) identifies four functions of a sponge city: (1) to coordinate urbanization and environmental resources; (2) to make the city adaptable to climate changes and natural disasters; (3) to change encourage reliance on the natural environment and the use of the ecological infrastructure to absorb, save, store, clean, use, and recycle rainwater; and (4) to ensure that landscapes' hydrological characteristics are unchanged after urban development. Che et al. (2015) defined the third type of sponge city from the perspective of LID. Narrowly construed, LID practices include gardens, green roofs, and roads made of permeable pavement. Broadly construed, a sponge city combines wetlands and floodplains with flood controls, pollution reduction, and the protection of water resources (Che et al., 2015).

In summary, there are three elements of a sponge city: the protection of original water resources; LID practices; and recycling water resources. In fact, the key concept of a sponge city is ecological compensation and the nature of return. Sponge City Initiative is based on a city's system, and is premised on the protection and repair of nature's protective mechanisms, combines civil engineering, hydrology, landscape and urban planning, and with the benefit of many types of technologies, improves the environment, facilitates cooperative urban flood management and development, and improves the city's health. The prevention of urban flooding is just one of a sponge city's many functions; the city needs the ecological infrastructure to absorb, store, and drain the rainwater during a storm.

China is in the early stages of urban flood management. From the perspective of governance, China's government is a state-centric model, and ignores the actions of non-state partners, as well as those of the residents who are the real defenders

[4] https://www.newsecuritybeat.org/2017/07/sponge-city-solutions-chinas-thirsty-flooded-cities/.

[5] http://chinawaterrisk.org/resources/analysis-reviews/sponge-cities-an-answer-to-floods/.

against urban flooding. Therefore, China's governance of urban flooding management is passive and inefficient. Under this type of governance, partners in urban flood management are inconstant, even across government departments. Further, since urban flood management involves urban planning, civil engineering, water conservancy, and flood-controls, if any one of these sections withdraws, the efficiency and functioning of urban flood management will be compromised. Shen (2013) has observed that although China has made great progress in the technologies used for fighting natural disasters, management technology is still backward, and there remain disparities among management activities in developed countries. Zhang and Lou (2014) wrote that public participation in urban flood management still has a long way to go, and the information available is asymmetric. Hou (2015) observed that there are problems with the emergency management of urban flooding. Ye (2008) described administrative principles and procedures as inefficient, autonomous, and uncoordinated.

To conclude, the improvement activities now underway in developing countries, and especially in Mainland China, stress infrastructure improvements and construction. Scholars and governments are beginning to emphasize ecological wisdom along with the Sponge City Initiative. However, urban management on urban flooding mitigation is still weak in Mainland China. Shenzhen is a rapid urbanization city attracting lots of peasant workers and tenants from China. As a garden city with a greening rate that is higher than in other Chinese cities, Shenzhen is also a new city with an improved sewage system design that separates rain and wastewater. Nonetheless, Shenzhen is still subject to floods. Now, Shenzhen is a city at the first construction stage of Sponge City Initiative. Therefore, Shenzhen is selected to do the research, with the purpose to identify the whether the present strategic measure of Sponge City Initiative could be used to reduce Shenzhen's losses from urban floods.

3.5 Risks, Threats, and Responses to Urban Flooding: Shenzhen

Shenzhen (N 22°27′–22°52′, E 113°46′–114°37′) is a major city located in the center of Guangdong Province, on the east side of the Zhujiang River. It borders Huizhou and Dongguan to the north, the Shenzhen River, Kowloon, and the New Territories of Hong Kong to the south, Daya Bay to the east, Lingdingyang Frith of the Pearl River Estuary to the west, and the cities of Zhuhai and Zhongshan are across the sea (Fig. 3.1).

Shenzhen has experienced rapid urbanization since China's "reform and opening" policy was established in late 1979. The former small village was then designated a special economic zone (SEZ). During the 1990s and 2000s, it was one of the world's fastest-growing cities. Shenzhen has six old districts (Futian, Luohu, Nanshan, Yantian, Bao'an and Longgang), and four new ones (Guangming, Pingshan, Longhua and Dapeng) (Fig. 3.2). There are 55 neighborhoods (excluding four

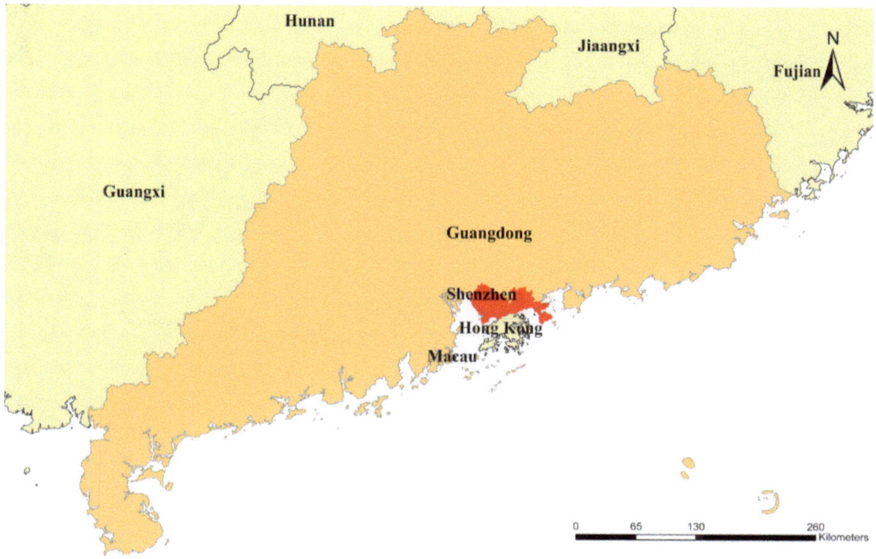

Fig. 3.1 Location of Shenzhen. *Source* Author

islands), and 792 communities, with a total land area of 1996.85 km^2. As one of China's top ten most livable cities, at the end of 2014 Shenzhen's population was 1077.89 (10,000 persons). Experts expect the number of population to increase to 3000 (10,000 persons) in the short future because of Shenzhen's attractiveness to immigrants.

3.5.1 Disaster Sources of Urban Flooding in Shenzhen

Shenzhen is built on a hilly area, and the city is higher in the east than in the west (Fig. 3.3). It has a mild, subtropical maritime climate, with plenty of rainfall and sunshine. From 1981 to 2010, Shenzhen's mean annual precipitation was 1935.8 mm,[6] which is characteristic of rainy regions. Except for the seasonal precipitation, Shenzhen and the Pearl River Basin get extremely heavy rain more frequent (Liu et al., 2016).

Some scholars have reported an uneven spatial distribution of the heavy rains in Shenzhen (Li et al., 2014; Song et al., 2008). They usually occur in the south-center of Shenzhen, including the southeast of Nanshan District, Futian District, west-center of Luohu District, and southeast Bao'an District; the southeast of Shenzhen, including Yantian District and the southwest of Longgang District; and east of Shenzhen, including Kuichong and Pingshan New District.

[6] http://www.szmb.gov.cn/article/QiHouYeWu/qihouxinxigongxiang/GaiKuangSiJiTeZheng/.

Fig. 3.2 Spatial distribution of the districts of Shenzhen. *Source* Author

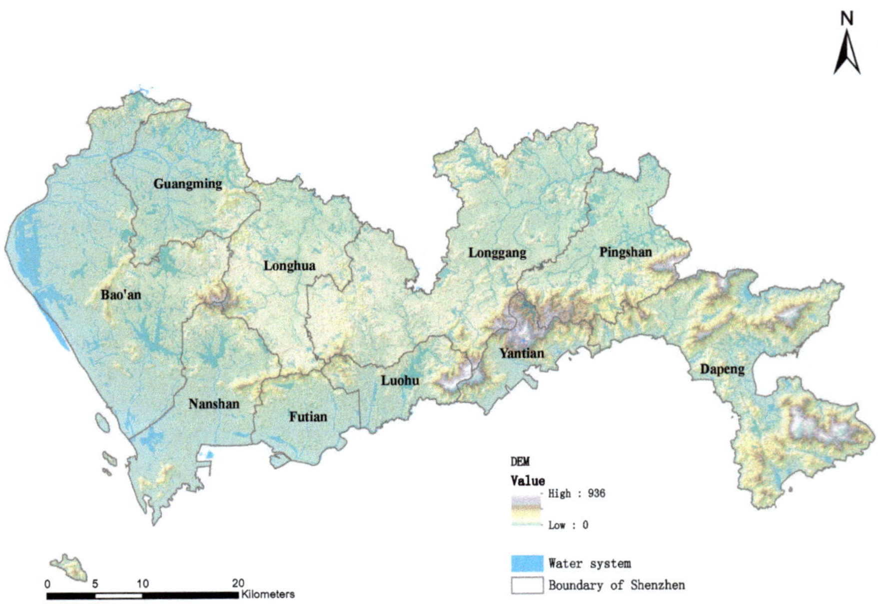

Fig. 3.3 Digital elevation model (DEM) of Shenzhen. *Source* Author

Statistical data[7] for Shenzhen from 2000 to 2007 reveal that typhoons occurred from July to September, and in these seven years, there were nine disasters, which led to 142 deaths, at an economic loss of CNY 2.77 billion. In 2014, eight tropical cyclones touched down within 500 km of Shenzhen area. Since 2012, there were more than 40 typhoon warnings in Shenzhen. Heavy rain is the main cause of urban flooding. With more frequent heavy and extreme-heavy rains, dealing with the problems associated with heavy rain has become urgent.

3.5.2 Rapid Development of Shenzhen

Since its establishment as a special economic region in 1979, Shenzhen has experienced rapid urbanization and economic development. Its astonishing development has made it famous around the world. Its yearly GDP has increased more than 8000fold, from CYN 0.2 billion in 1979, to CNY 1600.2 billion in 2014 (Shenzhen Statistical Yearbook, 2015), and it has an average annual growth rate of 23.5% (Shenzhen Statistical Yearbook, 2015, p. 23) (Fig. 3.4). In 2015, Shenzhen's GDP was lower only than the GDP in Shanghai, Beijing, and Guangzhou, however Shenzhen has the highest year-over-year growth rate, and the highest per capita GDP. Meanwhile, the tertiary industry/services sector in Shenzhen has seen a substantial increase, and the industrial structure tends to be rational, and based on high-tech and logistical industries. Financial industries have become the pillar industry, which demonstrates the strength of urban development. Over the past three decades, Shenzhen's population and population density soared. In 2014, the year-end permanent population of Shenzhen had climbed to 10.8 million persons, from 0.3 million persons in 1979, and in 2014 its population density was 5398 person/km^2[8] (Fig. 3.4).

In addition to the GDP and population changes, land-use changes are evident. During the past 40 years the amount of woodlands, croplands, wetlands, and unused land has decreased significantly, while urban areas have expanded. Figures 3.5 and 3.6 depict the dramatic land-use changes in Shenzhen from 2000 to 2010. In that decade, the city's green areas and water areas shrank as its road networks became denser.

3.5.3 Urban Flooding in the Context of Shenzhen

Considering both Shenzhen's present development situation and its increasing vulnerability to disasters, heavy rain and urban flooding have been occurring with increasing frequency.

---

[7] https://www.accuweather.com/en/cn/shenzhen/58194/weather-forecast/58194.

[8] Shenzhen Statistical Yearbook, 2015 (http://www.sztj.gov.cn/xxgk/tjsj/tjnj/201606/W02016060 7351981905550.pdf), p. 3.

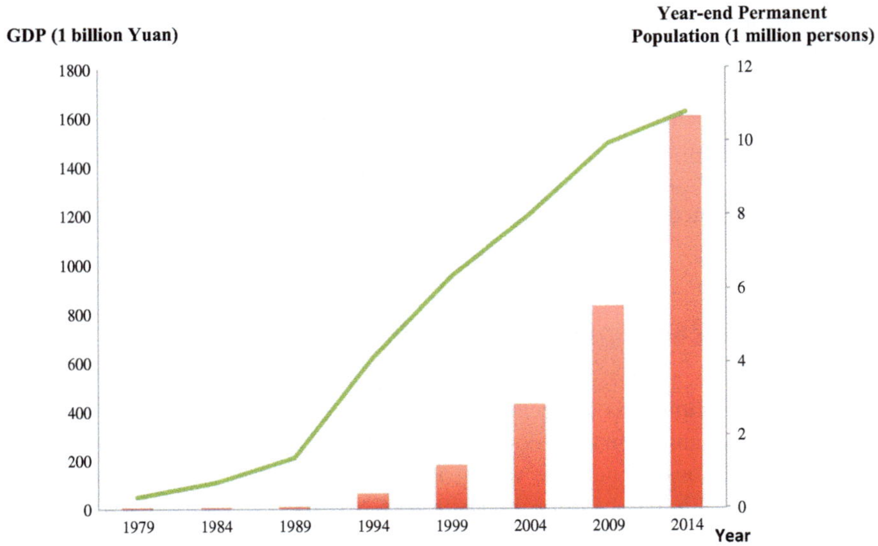

Fig. 3.4 GDP and the population of Shenzhen City. *Source* Author

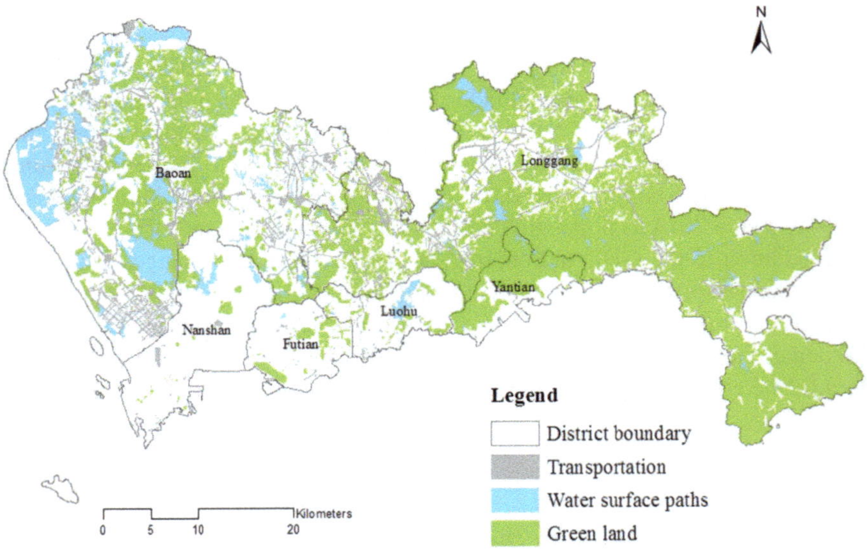

Fig. 3.5 Land use map of Shenzhen (2000). *Source* Author

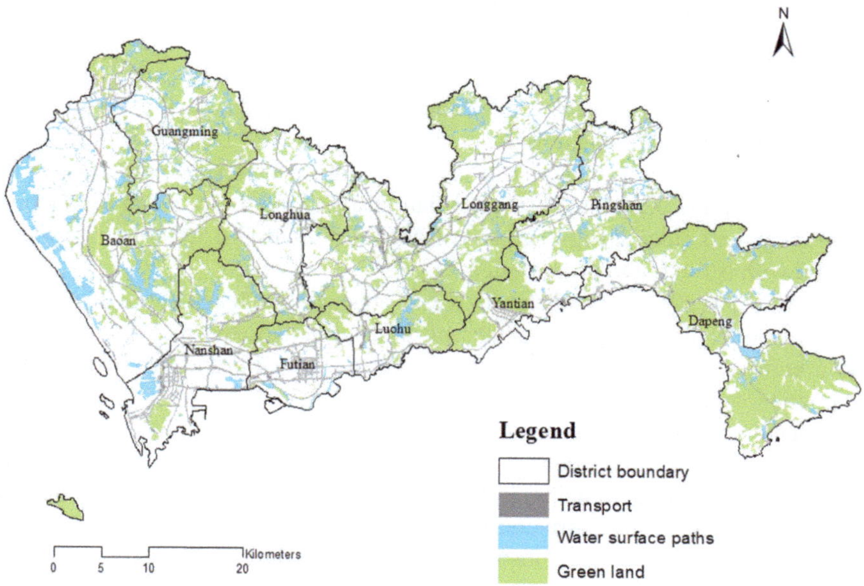

Fig. 3.6 Land use map of Shenzhen (2010). *Source* Author

Since 2007, there has been a great increase in the number of urban floods in Shenzhen, and they have led to huge economic and human losses.[9] On June 13, 2008,[10] heavy fain caused major flooding. Bao'an and Longgang, suffered a once-in-a-century rain, and because they are close to the sea, the seawater recharge turned the urban flood into a disaster. Eight residents died, six were missing, and more than 100,000 residents were displaced. More than 1000 buildings had been affected by the flooding at a loss of about CNY 1.2 billion.

On May 26, 2013,[11] Shenzhen had intensive heavy rain especially in Bao'an District. The heavy rain led to 60 places seriously flooded in Xixiang and Xin'an of Bao'an District, as well as dozens of cars flooded, which influence the travelling of residents within the city. Besides, the heavy rain also contributed to the flight cancellations. On August 30, 2013, another heavy rain attacked Shenzhen. Bantian of Longgang District is the most serious flooded place. Unfortunately, the heavy rain led to 2 people lost their lives, and more than 100 places flooded within the city.

On May 11, 2014,[12] another heavy rain attacked Shenzhen. Because of the heavy rain, newly-built airport of Shenzhen appear a major water leakage on its ceiling. Besides, the heavy rain also led the parking areas of airport appear flooding. Some roads of Shenzhen become the "river channels" because of the heavy rain and there

[9] News were searched on the platform of Google.

[10] http://www.chinadaily.com.cn/china/2017-06/14/content_29732678.htm.

[11] http://slide.news.sina.com.cn/c/slide_1_2841_32517.html#p=1.

[12] http://english.sina.com/china/p/2014/0511/699310.html.

are 200 urban floods, which lead to 2 people, lose their lives, and lots of motor vehicles stalling.

In May 2015,[13] there were continuously heavy rain came to Shenzhen. Heavy rain led to more than 70 places flooded and some road sections such as Shiyan Road of Bao'an District and Fengze lake tunnel of Longhua New District were seriously flooded with some cars anchored in the deep rainwater. Besides, the crossing of Lianhua Road and Xinzhou Road of Futian District, and the crossing of Nanping expressway and Huanggang overpass of Meilin, Futian District are all seriously flooded with 0.2 to 0.7 m of rainwater.

On May 10, 2016,[14] another heavy rain attacked Shenzhen. This time, about 50 places were flooded in Shenzhen and a canal collapsed which brought great inconvenience to residents, including travelling and daily life, and also brought huge economic loss to some industries.

In summary, the frequent of urban flooding is more and more these years in Shenzhen, and it is quite often to see and suffer urban flooding in Shenzhen. For the impacts from urban flooding, it does not lead to economic loss, such as the real estate, supply chain, transportation/logistics and financial services sectors may be affected by urban flooding, but also threatens the life safety of residents.

3.5.4 Current Situation and Issues of Urban Management on Flood Mitigation

Low design standards of the old city's drainage system were borrowed from the Soviet Union, which experienced less rain in the 1980s. In the intervening 30 years, Shenzhen has constructed more than 13,700 km of new drainage pipes.[15] However, these improvements in the city's infrastructure have not been enough to manage the heavy rainfall.

In 2016, Shenzhen governments invested CNY 1.3 billion[16] in containing urban flooding. This money will be used to improve drainage systems and build new ones, especially in the city's old community zones and the urban villages in Bao'an District and Longhua New District.

For sewage system management, Shenzhen is adopting some designs and planning. It is improving the design and return period for urban storm-drainage systems to mitigate the influences of urban flooding. Another strategy is modify the technologies to coordinate the drainage system's design and related special planning. Other plans are to designing integrated management measures based on LID; and to

---

[13] http://m.news.cntv.cn/2015/05/24/ARTI1432436788595312.shtml.

[14] http://www.szhyst8.com/a/xingyezhishi/2016/0511/250.html.

[15] http://www.szwrb.gov.cn.

[16] http://www.szwrb.gov.cn/ztzl_78021/ndmsss/yldzl/mtbd_87195/201612/t20161201_5516855.htm.

reset the storm intensity formula using modern models to calculate the flow rates of rainfall designs, which will increase the accuracy of the early warning of heavy rain.

Based on the learning from more advanced countries on the urban flood management, such as LID from America, SUDS from England, WSUD from Australia, in China, the government initiates the development of sponge cities. Several pilot sponge cities are now under development. Shenzhen has built a special institutional organization[17] for Sponge City Initiative led by the Shenzhen Municipal People's Government, including the Municipal Development and Reform Commission, the Innovation and Technology Commission, the Finance Commission, Urban Planning, Land and Resources Commission, the Human Settlements and Environment Commission, the Transport Commission, the Housing and Construction Bureau, the Water Bureau, the Urban Management Bureau, the Weather Bureau, the Quality and Technical Supervision Bureau, all the district governments (management committee) of Shenzhen Municipality, and the Shenzhen-Hong Kong Modern Service Industry Cooperation Zone of Qianhai, Shenzhen. Shenzhen has completed the planning and design of a sponge city. Four sponge bodies have been completed in Guangming New District, two are in progress and another 44 will be completed between 2016 and 2020.

3.6 Conclusion

Recent scientific research finds that climate change is more likely to increase rainfall extremes and urban floods. Cities that are undergoing rapid urbanization will experience increased vulnerability to severe flooding. Therefore, it is necessary to improve the capacity of residents to fight urban flooding. As a young, energetic, and inclusive city, Shenzhen attracts people of different educational backgrounds, incomes, and social positions. Its coastal geographical and climatic conditions are responsible for many rainy days every year, so urban flooding is always a concern.

Shenzhen is used as a case city to assess whether the can be used to change the present situation in Shenzhen for three reasons. Shenzhen experiences heavy rains (usually brought by typhoons) that lead to urban flooding. The dramatic changes in land use due to the Chinese government's political support are a natural test for dealing with urban flooding for a city under rapid urbanization. Shenzhen has undertaken strategic planning to design the construction of a sponge city. The Sponge City Initiative of Shenzhen can reflect the newest management concepts and measures, and the effectiveness of the measure is valuable to be explored.

[17] http://www.szwrb.gov.cn/ztzl_78021/bmzdgz/hmcsjs/.

Chapter 4
Research Methodology

4.1 Introduction

As discussed in Chap. 1, this study explores urban management on resilience improvement based on the evaluation of individual resilience, the identification of the determinants of individual resilience and the present strategic initiative for nurturing individual resilience. Three groups of research questions are thus posed: (1) What was the current performance of individual resilience of residents vulnerable to urban flooding? And did the individual resilience vary among different communities? (2) What were the determinants (People-Disaster-Place-Society) of individual resilience? (3) Whether the strategic measure of Sponge City Initiative could help nurture individual resilience? And what were the place-responsive measures for individual resilience nurturing?

Building upon the theoretical foundation and conceptual framework presented in Chap. 2 and the empirical foundation presented in Chap. 3, this chapter presents the analytical framework and research methods to answer the three groups of research questions. This Chapter comprises two parts. The first part explains a three-level analytical framework and research approach, including the individual resilience evaluation and variations of individual resilience among different communities; the identification of the limiting factors of individual resilience; and the exploration on the effectiveness of the strategic measure of Sponge City Initiative in nurturing individual resilience at the local scale and the possible place-responsive measures to address the potential loopholes. The second part introduces the research methods, including reasons for research area selection, data collection and data analysis methods.

J. Song, *Individual Resilience to Urban Flooding and the Implications for Urban Management*, https://doi.org/10.1007/978-981-97-0546-7_4

4.2 Analytical Framework

In line with the three groups of research questions, Fig. 4.1 illustrates the analytical framework of the research. The conceptual relationships among them and the research hypotheses were developed in Chap. 2. The primary task is to evaluate the individual resilience of the selected research area and provide, based on the influencing factors identified and the judgment on the effectiveness of Sponge City Initiative for resilience improvement, place-responsive measures to nurture individual resilience. Different from similar researches that emphasize the objective attributes of different individuals and around, the study also focuses on collecting data on the subjective personal perceptions of around as the key determinants of individual resilience. These factors are comprised of personal perceptions (People) of urban flooding (Disasters), quality of physical environment (Place) and social environment (Society). Therefore, the interactions among people, disaster, place and society are carefully explored for their potential influence on individual resilience to urban flooding, providing references and information for further recommendations on individual resilience improvement for the vulnerable social group to urban flooding.

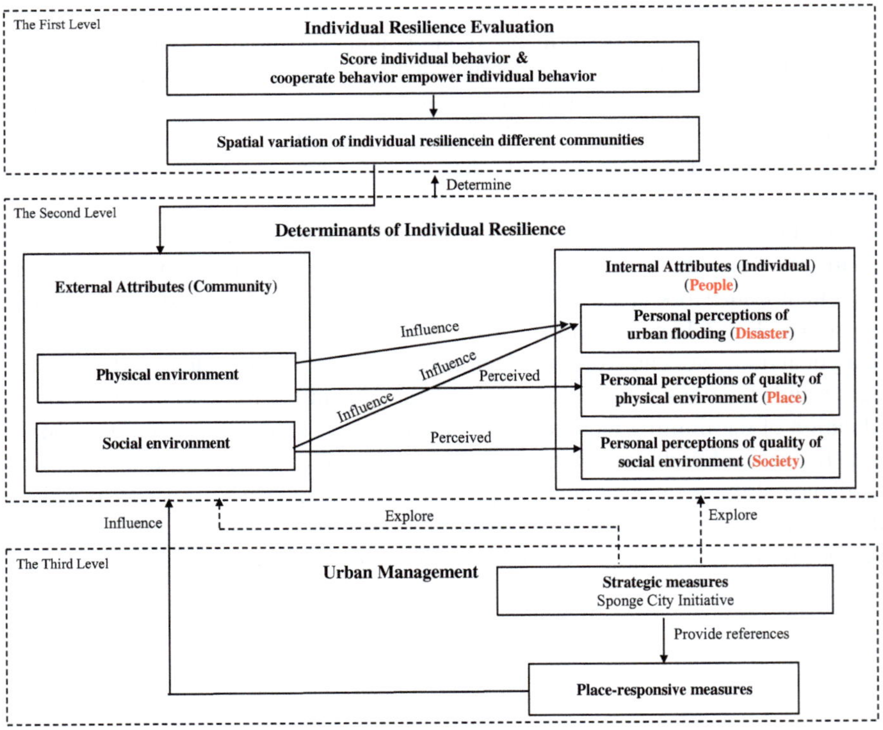

Fig. 4.1 Analytical framework. *Source* Author

The first level has two aims. (a) Individual resilience is evaluated. Individual resilience is scored and evaluated on a residence location basis through the coping and adaptive strategies adopted by individuals as well as the ones adopted by the corporate actors, i.e. managers of living communities empower individual behavior. (b) Individual resilience variation is identified by exploring whether residents of different communities demonstrate commonalities in their individual resilience performance. This part of the research is based on the first hypothesis: There are significant differences in individual resilience to urban flooding for residents in different communities.

The second level identifies the influencing factors of individual resilience. Both external environment and internal factors of individuals as potential determinants of individual resilience are explored. Since individual resilience is evaluated on a residence location basis, external environment of individuals mainly includes the physical and social environment of living communities of individuals. Communities with different physical environments and social environment levels are identified to explore their degree of individual resilience. Internal attributes of individuals as potential determinants of individual resilience are also explored. Except the socio-demographic attributes of individuals, in this research, creatively, factors demonstrating the personal perceptions of urban flooding, the quality of physical environment and social environment are collected as the potential determinants of individual resilience. This part of research is based on the second hypothesis: Individual perceptions of urban flooding, living place and living society can influence individual resilience.

The third level identifies the institutional arrangements for Sponge City Initiative, and explores if the Sponge City Initiative can nurture individual resilience. A main focus of this part is that place-responsive measures are provided as the necessary means to cultivate individual resilience. Place-responsive management measures are provided based on the experiences and lessons drawn from Sponge City Initiative. This part of research is based on the third hypothesis: Individual resilience cannot be nurtured through resilience-oriented measure, i.e. Sponge City Initiate from central government of China.

Combining the analytical framework in Fig. 4.1, the research approach to test research hypotheses and answer research questions is shown in Fig. 4.2.

4.2.1 Spatial Variation of Individual Resilience

Individual resilience evaluation is the first step for this research. Theory of action involves corporate actors at the societal level and persons at individual level. Facing disturbance, human behavior demonstrating the capacity for coping and adaptation does not only include the behavior adopted by individuals, but the ones adopted by corporate actors, which can empower individual behavior (Coleman, 1986). In this research, all the residence location-based resilience-oriented individual behaviors in terms of urban flooding are identified based on interviews with governments,

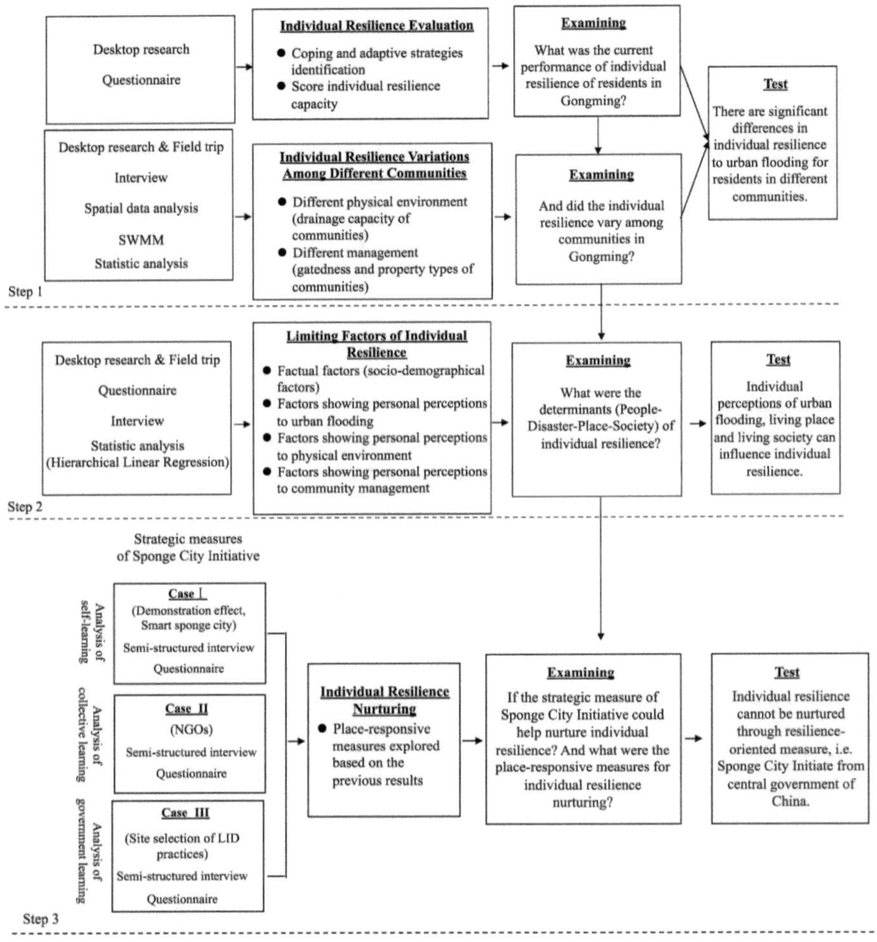

Fig. 4.2 Research approach. *Source* Author

scholars, property managers and residents. These behaviors can be divided into the following groups: (1) preparing for the potential urban flooding risk; (2) learnt to cope with and adapt to urban flooding; (3) the past coping strategies during urban flooding; (4) habits to cope with and adapt to urban flooding; and (5) the creative and flexible strategies to adapt to urban flooding (Table 4.1).

Then, a questionnaire is designed based on the resilience-oriented behavior at individual level on a residence basis. Randomly selected individuals undertook self-report on whether they and their living communities have adopted these strategies. With these responses, the resilience performance of individuals can be scored for evaluation.

After the calculation of individual resilience for the surveyed residents, the average score of individual resilience of 11 communities can thus be calculated. Then

Table 4.1 Questions to get self-reported behaviors for individual resilience evaluation

Questions
(1) *Preparing for the potential urban flooding risk*
1. Do you do some preparation for heavy rain or urban flooding?
2. Does your living community have the daily maintenance about the pipelines and pipe junctions?
3. Does your living community have commissioners to repair the pipelines after they have been jammed?
(2) *Learnt to cope with and adapt to urban flooding*
4. Do you learn how to help yourself escape from urban flooding?
5. Do you take the initiative to study the knowledge of disaster prevention and risk aversion?
6. Does your living community provide education for improving your capacity for adapting to disasters?
(3) *Past coping strategies during urban flooding*
7. Do you know where can you escape from the heavy rain or urban flooding when you are outside?
8. Do you know which areas usually experience urban flooding around your living communities?
9. Does your living community have some measurements to overcome the urban flooding?
(4) *Habits to cope with and adapt to urban flooding*
10. Do you have the habit to check weather report?
11. Do you disseminate safe-haven news to your families and friends after you have known it?
12. Do you know the early-warning signal of heavy rain?
(5) *The creative and flexible strategies to adapt to urban flooding*
13. Do you follow the early-warning signal to arrange your travel flexibly?
14. Do you participate into the new strategies building for disaster prevention of your living community?
15. Does your living community have some innovative strategies for adapting to disasters?

Source Author

the spatial distribution and characteristics of individual resilience among different communities could be identified and explored.

4.2.2 Exploration of the Determinants to Individual Resilience

The review of individual resilience performance has revealed its association to the external living environment of individuals, such as the physical environment and the social environment of human communities (Cutter et al., 2008; Gunderson, 2010).

In the physical environment, the natural and manmade infrastructure plays a critical role in coping with urban flooding. Natural ecological systems and green land can absorb and transfer rainwater as well as reduce the pressure of drainage system to absorb rainwater at times of heavy rain. The manmade drainage system is based on advanced technologies and can help an impervious surface absorb rainwater (Bruneau et al., 2003). In the event of urban flooding, the most relevant and vital urban infrastructure is the sewage system. Sewerage systems can "absorb, adapt, and recover from the effects of a disruptive event while attempting to continue delivery of critical infrastructure services" (Vugrin & Camphouse, 2011). It is concluded that individual resilience to urban flooding has a close relationship with the absorption capacity of drainage systems (Ahern, 2011).

There are some attributes concerning social environment of communities that further influence resilience. Property types, property rights and community structure are three of the key elements in identifying the differences of social environment between communities (Dolan & Walker, 2004; Naumann et al., 2009; Yohe & Tol, 2002). Social factors, such as education level, poverty, well-being of communities and crime rates are important factors of the social environment (Arouri et al., 2015; Ebersöhn, 2017; Friend & Moench, 2013; Hess et al., 2012). Besides, the economic disadvantage, employment, and spatial distribution of sporting clubs, service organizations, hospitals, and schools reflect the social attributes of communities (Maybery et al., 2009). Furthermore, neighborhood quality, neighborhood type, gatedness, safety and management play important roles in distinguishing different types of communities (Brown & Westaway, 2011; Dixon & Dupuis, 2003; Godschalk, 2003; Li et al., 2012; Wickes et al., 2015; Youssef, 2015). Moreover, the management of people, organizations, and resources is important in influencing social environment of communities (Colussi, 1999). In summary, poverty, property type, property management, neighborhood safety, gatedness and crime are the main factors distinguishing a community from others.

In this research, attributes distinguishing different environments of communities are identified through desktop research, field trip observation, and interviews. Among the physical factors influencing individual resilience to urban flooding events, the efficiency of sewage system is selected to distinguish different communities. The efficiency of sewage systems, named drainage capacity of sewage systems, does not only depend on the density and length of the sewage system, but also land use, ecological land density, slope and elevation of the location of sewage systems. Therefore, it is suitable to select drainage capacity to reflect a community's physical environment. Among the factors reflecting the social environment of different communities, "gatedness of the community" and "property types" are selected to identify their influence individual resilience to urban flooding as most other factors concerning the social environment of community could be deduced from these two factors (e.g. residents of non-gated urban villages usually have lower income levels and higher crime rates).

The external attributes of community social environment assumed to influence individual resilience are shown in Table 4.2.

Table 4.2 External environmental factors of community

Category	Subcategory	Variables
Factual factors	Physical environment Social environment	Drainage capacity of sewage system Gatedness of the community Property types

Source Author

Internal attributes on individual level, literature about urban flooding resilience has revealed some common attributes of individuals can influence and determine human behavior and distinguish between those with high and low social resilience, such as age, gender, income, education level, and ethnic group (Cohen et al., 2016; Leykin et al., 2013; Soetanto et al., 2016).

Besides, as discussed in Chap. 2, personal perceptions are assumed to be the potential determinants of individual resilience. In this research, three types of such determinants are obtained based on the literature (Alshehri et al., 2015; Burnham & Ma, 2017; Burton, 1993; Cohen et al., 2016; Dolan & Walker, 2004; Horton & Wallander, 2001; Leykin et al., 2013; Soetanto et al., 2016; Yohe & Tol, 2002). The first type is the factors reflecting residents' perceptions of urban flooding, such as considering it dangerous, remaining sensibly prepared to deal with it, personal efficacy and social support associated urban flooding as well as the previous experiences of urban flooding. The second type consists of factors reflecting individual perceptions of the quality of physical environment of their living places. These include the satisfaction on the environment, garbage disposal, and the amount of construction land and ecological land in the community. The third type consists of factors reflecting individual perceptions of social environment, including factors demonstrating individuals' perception on social cohesion (Table 4.3).

All the potential determinants of individual resilience on individual level were obtained from individual self-reporting through questionnaire survey. Socio-demographic factors, such as gender, age, income and so on work as the control variables in this research. As mentioned above, the objective attributes of community environment can influence individual resilience to some degree. In order better to understand the influence of subjective attributes of individuals on individual resilience, Hierarchical Linear Model (HLM) is conducted to identify the final determinants of the individual resilience in the selected research area. HLM can help to identify the determinants on the individual level and the different influences of its higher level, in this research, the level of their communities, which better reflect the determinants of resilience performance of the special population group in the selected research area (Loo et al., 2017).

Table 4.3 Socio-demographic and subjective attributes of individuals

Category	Subcategory	Variables
Socio-demographic factors	Control variables	1. Gender 2. Marital status 3. Age 4. Income 5. Occupation 6. Education level
Perceptions of urban flooding	Personal efficacy (PE)	1. Leadership 2. Ability to lead others to escape from hazard during urban flooding 3. Public demands on related urban planning and management
	Personal risk perception (PRP)	4. Feelings about urban flooding 5. Sensitive to urban flooding 6. Concerns about children during urban flooding 7. Concerns about the elderly during urban flooding
	Social support perception (SSP)	8. Help and support from the government 9. Help and support from their workplaces
Perceptions of urban flooding	Previous experiences of urban flooding (PEOF)	10. Number of days worked per week 11. The primary means of commuting 12. Experienced urban flooding 13. Previous losses during urban flooding
Perceptions of quality of physical environment	Physical environment perceptions (PEP)	14. Amount of construction projects in progress around communities 15. Areas of ecological lands around communities 16. Living environment of communities 17. Waste management created by natural hazards of communities
Perceptions of quality of social environment	Social cohesion perception (SCP)	18. Mutual trust during times of difficulty 19. Mutual help during times of difficulty 20. Perception of life in your residential area 21. *Hukou* status 22. Housing ownership

Source Author

4.2.3 Place-Responsive Measures for Individual Resilience Improvement

Social learning theory can connect resilience nurturing (positive resilience performance changing) to urban management. The learning of the government is particularly important, for this learning directly influences and determines the government's behavior and attitudes in decision-making on the measures for resilience nurturing, which may transform residents' present behavior towards more resilience-oriented ones (Armitage, 2006; Smith & Stiring, 2010; Tran & Shaw, 2007). Therefore, based on the social learning theory, the possible problems associated with individual and collective learning in relation to resilience can be identified. Governments should offer resilience nurturing as a public service. Therefore, the third step of this research is to explore if the resilience-oriented institutional measures established by governments can facilitate individual resilience improvement. In this research, the selected resilience-oriented strategic measure is "Sponge City Initiative". This initiative includes several institutional arrangements such as appointing a special working group and undertaking dedicated planning exercises. The focused measures of Sponge City Initiative include the spatial and temporal arrangements for the upcoming projects of sponge city construction and social services during sponge city development.

More specifically, this study selected the following measures of Sponge City Initiative to check whether they help resilience nurturing. First, the visits to the completed LID projects and the demonstration effect generated, for it may help residents achieve self-learning to cultivate resilience. Second, Smart Sponge City associated with projects under the Sponge City Initiative, as the cooperation between the government and technology companies may help enhance information-sharing networks for better collective learning about resilient behavior. Third, redevelopment projects incorporating LID practices in construction and site selection, because they can reflect the quality of government learning on resilience nurturing by evaluating whether there is spatial dislocation for LID practices.

Political decisions for resilience improvement are required on a locality-specific basis, and the suggestions and measures provided by policy-makers should be place-responsive (Adger, 2003). Therefore, finally, localized measures for nurturing of individual resilience are provided based on the identification of present learning problems, the effectiveness of current government strategic measure and the interactions among people-disaster-place-society.

4.3 Research Method

4.3.1 Study Area Selection

Considering its special natural, socioeconomic, and institutional conditions, Gongming, a densely built sub-district in Guangming New District of Shenzhen located in the Maozhou River Basin is the ideal candidate as the study case.

Guangming New District located in the Maozhou River Basin, consists of two sub-districts, Gongming and Guangming, with the latter comprised mostly of farmland. As, this study focuses on urban flooding, Gongming, with a larger amount of built-up area, is selected for analysis. With 19 communities and a total area of 100.3 km^2, Gongming has a subtropical oceanic climate with an annual average temperature of 22 °C and abundant rainfall. Maozhou River and its tributaries cross the Gongming neighborhood from northwest to southeast. It is a hilly area with low mountains, and the Liantang, E'jing, Tiekeng, and Gongming reservoirs are all located within the sub-district.

This study is primarily focused on urban areas with high population densities. Based on this criterion, a study area located in the southwestern part of Maozhou River is identified (Fig. 4.3), covering 11 of the 19 communities in Gongming Sub-district. The names of the communities are listed in Table 4.4.

Gongming covers 37.68 km^2 and the built-up area within the study area is 26.31 km^2, which accounts for 69.8% of the study area. Industrial and residential lands are the main components of the built-up area and development activities are quite intensive in this area. Based on the statutory plan map of Guangming New District, small property housing (urban villages) accounts for 80.11% of its residential zones, with most of them being rental flats for peasant workers.

Guangming New District is a pilot district for Sponge City Initiative in Shenzhen. Sponge City Initiative is a creative measure of the central government of China to improve the resilience of the country against natural disasters, especially for drought and urban flooding. A dedicated plan[1] for Sponge City Initiative in Shenzhen's Guangming New District proposes a total of 51 projects with a total investment of CNY 5.038 billion in five years (2016–2020). These will be financed by the government and by PPP, with a key emphasis on strengthening the sewage systems through collaboration between the government and the industry.

Most of the 51 planned projects are mostly clustered around Gongyuan Road, Guanguang Road, sports center of Guangming New District, and Fenghuangcheng, a new town of Guangming New District under construction. Given the depreciated sewage systems, revamp and maintenance should presumably be a key part of the program of Sponge City Initiative as well. Eight projects of this kind are found in Gongming (Table 4.5).

In summary, Gongming is an urban sub-district of Shenzhen on its way to rapid development, despite it is being the community with almost the lowest GDP in

[1] *Source*: Shenzhen government.

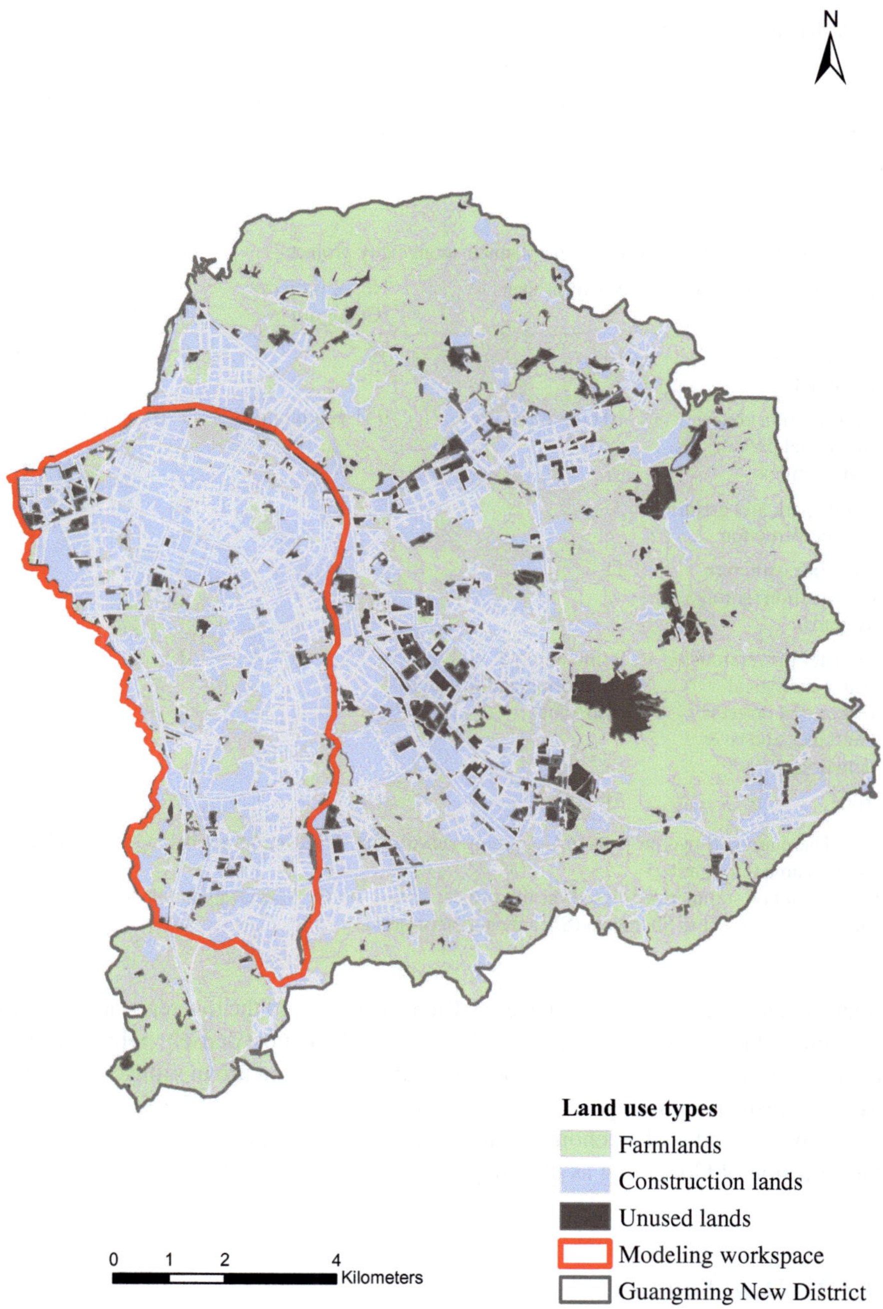

Fig. 4.3 The study area (Gongming). *Source* Author

Table 4.4 Names and affiliations of the surveyed communities (Gongming)

District	Sub-district	Community
Guangming New District	Gongming	Citianbu, Genzhuyuan, Gongming, Heshuikou, Hongxing, Jiangshi, Mashantou, Shangcun, Tianliao, Xiacun, Yulv

Source Author

Table 4.5 Spatial distributions of the planned sponge city projects

Project classification	No. of projects	No. of projects (Guangming)	No. of projects (Gongming)
Municipal roads construction	16	16	0
Construction zones (industrial park or communities)	10 (1 community)	9 (1 community)	1 (0 community)
Public park and green land construction	5	5	0
Civil infrastructure construction (sewage systems)	11	7	4
Disaster prevention and water ecological restoration projects (wetlands and river networks)	9 (including 2 monitoring and research platforms)	4	3
Total	51	41	8

Note There are still two projects in Gongming sub-district, however do not belong to the selected research area (Gongming)
Source Author summarized based on the project library on the special planning for sponge city construction project in Shenzhen (Shenzhen government)

Shenzhen at the urban sub-district level. But just because of the low rents, it has led to the influx of rural migrant workers, making it one of the most densely populated sub-districts of Shenzhen with 80% of its residential land being urban villages. Selected as a pilot district for Sponge City Initiative, spatial analyses reveal that Gongming has high physical and socioeconomic vulnerabilities to urban flooding. All these justify the selection of Gongming as the study area for the purpose of this research.

4.3.2 Data Sources

Both quantitative and qualitative data were used in this research. In line with three groups of research questions, five data sets were collected (Table 4.6).

Table 4.6 Data sets and introduction

Data type	Data introduction	Source
Specific division	Spatial rainfall data on May 11, 2014, and rainstorm intensity formula (once-in-200-years) for Shenzhen (2015)	Website-Chinese climate data (data.cma.cn); Bureau of Meteorology, Shenzhen
Statistical yearbook	Population, per capita GDP, gender, and age of residents of Guangming New District, Gongming sub-district and Guangming sub-district	Yearbook of 2016, Shenzhen, The sixth census of Guangdong province (2010)
Socioeconomic information and resilience-related data	Gender, marital status; *hukou* status, income, age, education levels Questions comprised of evaluation of individual resilience performance Questions to get the potential determinants of the individual resilience	Questionnaire, interviews and field observations (2016.07.15–2016.10.30)
Remote sensing and GIS data	Land use map (2015), administrative map (2015), DEM, statutory plans of Shenzhen (2015), spatial distribution of soil types and catchments, spatial patterns of rainfall on 11, May, 2014, Shenzhen	Shenzhen governments
Ancillary data	Infiltration rate data of different materials, as well as the other variables applied in the SWMM	Reference to related research and materials

Source Author

The first dataset is the rainfall and rainstorm intensity data from Shenzhen government for the calculation of drainage capacity for each community. In this study, the rainfall on May 11, 2014 was selected for urban flooding simulation because the spatial distribution data of urban flooding were available for that event and can be used to verify the accuracy of the simulation results. The precipitation data was downloaded from the website of Shenzhen Meteorological Data System.[2] The rainstorm intensity formula (2015)[3] was adopted from the website of Meteorological Bureau of Shenzhen Municipality.[4] In this study, a 200-year storm is selected to simulate the flood-prone areas in Shenzhen.

[2] http://data.szmb.gov.cn.

[3] http://www.szmb.gov.cn/site/szmb/baoyuqiangdugongshi/index.html.

[4] http://www.szmb.gov.cn/article/ZhuanTi/byzt/.

The second type of data is from statistical yearbooks, which is applied to design the sample of questionnaire in my research area. The census data of the New District of Shenzhen are available in statistical yearbooks of different domains. The sixth census of Guangdong Province was conducted in 2010 and the spatial distribution of the population is assumed to have remained unchanged since then.

The third type of data is socio-demographic data, personal perception factors, as well as the strategies adopted by individuals to cope with and adapt to urban flooding from the questionnaires, interviews and field observations. They are used for the individual resilience calculation and explore the potential determinants of individual resilience.

Fourth, remote sensing images provide a regional view and detailed spatial coverage of the earth's surface. In this study, remote sensing images were used to identify the land use cover and digital elevation model (DEM) of the selected area. The administrative map (2015) was used to delineate the administrative boundaries of the cities, districts, and zones to facilitate the spatial analysis. The spatial data on road networks and water systems were collected from the land use map (2015).

Finally, some parameters applied to the SWMM are obtained to calculate drainage capacity for each community, which reference the related research.

4.3.3 Data Collection Methods

4.3.3.1 Questionnaire

A questionnaire was used to obtain (1) self-reported coping and adaptive strategies, (2) socio-demographic characteristics of local residents, and (3) the personal perceptions of urban flooding, quality of physical environment and social environment.

The survey was conducted from July 15 to August 2, 2016 in Gongming. Face-to-face interviews were also conducted to improve the understanding of the questions for the surveyed people.

Sampling design

The target population was adult residents ($\geq$ 16 years old) having lived in Gongming for longer than six months. Scholars consider men aged 16 years or above as the rational men, who can make choices by themselves independently. In this research, I need a self-reported questionnaire, 16 is the minimum requirements. For this research, I focus on the migrant workers and permanent residents. I want to exclude the ones who are temporal residing there, such as travelling or visiting friends or something like that. In Shenzhen, people who live at least six months can apply the residence permit. Modified convenience sampling was used because the population of the study area is too large to include every individual or to perform random sampling. Most respondents were selected randomly in public spaces such as public squares, parks, streets, shops, and bus stops.

Table 4.7 Main characteristics of population of Guangming New District

Types	Data
Age	$\geq$ 18 accounts for 96%
Gender	Male/Female = 1.22
Hukou status	Holder/Non-holder = 12%

Source The sixth national population census and statistical yearbook of Shenzhen in 2015

To reduce the drawbacks of convenience sampling, several methods were applied to increase the representativeness of the sample. The survey period includes both weekdays and weekends from 9:00 am to noon and from 3:00 pm to 8:00 pm, ensuring that both workers and non-workers are surveyed. Green and open spaces, retail stores, factories, bus stops as well as the entrance of the communities were selected to conduct the survey to include a variety of residents. Statistics on age, gender, and percentage of Shenzhen *hukou* holders were taken from the statistical data of the Guangming New District of Shenzhen yearbook 2015 and the sixth national population census (Table 4.7), which is referenced to design the questionnaire sample.

Questionnaire design

The questionnaire consists of four parts: (1) *personal and socioeconomic characteristics* to get personal information; (2) *understanding on urban flooding* to get personal perceptions of urban flooding and (3) *personal capacity* to get self-reported strategies to cope with and adapt to urban flooding; (4) *community/living zone prevention* to get self-reported strategies from corporate actors to cope with and adapt to urban flooding, personal perceptions of quality of physical environment and social environment. Some questions were based on a hypothetical scenario of a serious urban flooding event. Two types of information were collected by this questionnaire design: (a) factual questions on the socio-demographic features and the coping and adaptive strategies of individuals and the corporate actors, such as their communities to manage urban flooding and (b) attitude questions on the personal perceptions of urban flooding, quality of physical environment and social environment. For factual questions, single-choice questions and for attitude questions, multiple-choice or short questions are used as the main methods to collect information. Since hypothetical scenarios were presented to the respondents, they were not allowed to change their previous choices after receiving some new information on that question. For example, some respondents who answer "no" to the question "Do you think urban flooding is dangerous?" may want to change their answer after they hear the next question "Which of following situations do you think is the most dangerous in a hypothetical heavy urban flooding event?" Except for gender, marital status, and *hukou* status, most socio-demographics questions were used for comparison. A sample of the questionnaire is available in Appendix A.

To reduce the possibility of misinterpretation, triangulation (Denzin & Lincoln, 2002) was applied during the questionnaire survey. Secondary data sources were used to verify the accuracy of the primary data. For example, the 12th question in *personal capacity* part of questionnaire is to ask can you tell us your understanding of sponge city? And a testing questionnaire followed that is that please write down your understanding of it if you know it. It can help to verify if the responder gives a right answer. However, the disadvantages of self-reporting questionnaires—the possible misunderstanding the questions or the influence by scenario descriptions—cannot be fully removed still.

4.3.3.2 Semi-structured Interviews

Semi-structured interviews were conducted from September 6 to October 30, 2016. Based on a prepared outline, direct responses and other primary and secondary data were collected. The interviewees included local government officials, planners, developers, consultants, neighborhood committee members, and residents.

Information collected from the interviewees was used for qualitative analysis, focusing on the interactions among different actors in urban management for urban flooding mitigation. The responses of local residents were used to examine the effectiveness of urban management and the sensitivity of residents to changing conditions. Residential committee members and housing managers were interviewed to identify the measures to prepare, prevent, and manage urban flooding. The responses of developers, planners, scholars and NGOs were used to explore the efficiency of Sponge City Initiative for the residents' resilience improvement. The responses of government officials were used to identify the direct measures of urban management for urban flooding mitigation. Thematic method is applied to analyze the qualitative data.

The final sample included 38 interviewees: 16 residents, five developers, four government officials, four planners, two housing managers, two residential committee members, three staff members of NGOs and two scholars. The sample meets the requirements of quantitative research. The interview questions and list of interviewees are in Appendix B.

4.3.3.3 Site Observation

We performed site observations based on the questionnaires and interviews. The goal of site observations was to understand the living environment of the residents and ascertain the effectiveness of management by housing managers, residential committees, and governments. For example, the effectiveness of the management of sewage systems can be assessed by observing the conditions of sewage system units (Appendix D). Observation can also determine the gatedness of a community.

4.3.3.4 Desktop Research

Desktop research is important for collecting secondary data. Some measures and planning tools can be found on the Internet, including the institution of urban flood management in Shenzhen. The flood-prone areas and the people in charge of those areas can be identified on the websites. In this research, the Sponge City Initiative is a real-life ongoing management case analyzed, and its management committee, specific measures, and the progress of the ongoing projects can be downloaded from the website.[5] Planning documents, planning standards as well as some websites were consulted to identify the details of the urban flooding management and urban development schemes (Table 4.8).

4.3.4 Data Analysis Methods

4.3.4.1 Individual Resilience Assessment

Behavioral research has used self-reported measures to gauge the responses to surroundings, such as environmental changes (Chao & Lam, 2011; Lam & Cheng, 2002). More and more scholars choose self-reported questionnaires to get people's behaviors, and then score their self-reported choices through evaluation standards, especially in personal healthy research (Gionet & Godin, 1989; Hollon & Kendall, 1980; Lee & Podsakoff, 2003; Montano & Kasprzyk, 2015; Podsakoff et al., 2014). In this research, individual resilience was evaluated based on the self-reported coping and adaptive strategies of the residents and their living communities. The design of the questions in the questionnaires uses conceptual, theoretical, and empirical justification from previous research to represent individual resilience.

By referencing the methods for scoring the measure of quality of life and human health,[6] 15 questions were asked to obtain information on the respondents' coping and adaptive behavior. The questionnaires provide raw data that need to undergo a process of transformation and theoretical orientation. *Transformation* means the choices are converted into different variables with scores. This is necessary to classify all variables uniformly. The theoretical orientation of each variable adjusts the variables so that individual resilience has a theoretical meaning; larger values correspond theoretically to higher individual resilience levels. In this study, the higher the index value, the higher the resilience performance level of individuals. The resilience performance was scored on a two-point or three-point Likert scales (Table 4.9) based on a literature review (Galatzer-Levy et al., 2014). The average of all the selected strategies is the final score of individual resilience performance.

[5] http://www.szwrb.gov.cn/ztzl_78021/bmzdgz/hmcsjs/.

[6] http://www.measuringimpact.org/s4-sf-36.

Table 4.8 Supplementary materials for urban management analysis

Sources	Items
Planning documents	Flood prevention standard (GB50201-2014) Engineering design standard for urban flood prevention (GB/T50805-2012) Environmental quality standard for surface water (GB3838-2002) Code of urban wastewater engineering planning (GB50318-2000) Code for design of outdoor wastewater engineering (GB50014-2006) (2014 edition) Code for comprehensive planning of urban engineering pipeline (GB50289-98) Engineering technical code for rain utilization project with the community (GB 50400-2006) Guidelines for the planning of urban river system (SL431-2008) Code for planning the urban river system (GB50513-2009) Environmental quality standard of surface water in China (GB3838-2002) Water pollutant emission limits (DB 44/26-2001) Sponge city construction technology guide-low impact development and construction of storm water drainage systems (advance copy) Performance evaluation and assessment methods for sponge city construction (advance copy) Interim previsions on the special planning for sponge city (architectural program [2016] 50) Urban planning standards and guidelines revised for Shenzhen (edition 2013), Shenzhen government, 2014. 1 Low impact development of technical standard on comprehensive utilization technology of rainwater (SZDB/Z 145-2015) Technical code for rain utilization project (SZDB/Z 49-2011)
Planning standards	Shenzhen's economy and social development of the 13th Five-Year Plan, 2016.1 Shenzhen Municipal Government work report, 2016.1 Shenzhen statistical yearbook (2014–2016), 2015.3 Master plan of Shenzhen (2010–2020), 2010.10 Shenzhen city development strategies 2030, 2005.12 Shenzhen water strategy, 2007.8 Shenzhen city wastewater system planning revision (2011–2020), 2011.7 Shenzhen city drainage system planning (2007–2020), 2006.08 Shenzhen city revision and updating planning for flooding and tide prevention and river regulation planning (2014–2020), 2013.3 Shenzhen utilization system layout planning, 2011.1 Shenzhen blue line planning, 2008.6 Shenzhen drainage redevelopment and implementation planning, 2011.12 Shenzhen green land system planning, 2004.12 Other related planning and engineering design Water affaires development in Shenzhen in twelfth five-year plan, 2011.12 Shenzhen road network planning in twelfth five-year plan, 2011.12 Shenzhen urban planning "A picture" Immediate-term planning for training works in flood control and drainage of Shenzhen, 2014.11 Comprehensive plan for drainage (rainwater) flood prevention of Shenzhen city, 2016.1 Shenzhen water quality programme of work (2015–2020), 2015.12 Shenzhen green road network planning, 2011.9 Survey data for rainwater pipelines of the whole city, 2014.3 Reconnaissance information for drainage system of original outside the special administrative region, 2014.3 Reconnaissance information for drainage system of Shenzhen water supply group, 2014.3

(continued)

Table 4.8 (continued)

Sources	Items
Websites	www.sz.gov.cn (Shenzhen, China) www.szwrb.gov.cn (Shenzhen Water Authority) www.gd3f.gov.cn (wind, flood, and drought prevention headquarters of Guangdong) www.szmb.gov.cn (Meteorological Bureau of Shenzhen Municipality)

Source Author

4.3.4.2 Determinants of Individual Resilience Identification (Community Level)

As reviewed in Chap. 2, the literature suggests the attributes that distinguish different types of communities include the physical environment and social environment. Drainage capacity is calculated with SWMM (Appendix C) reflecting the physical environment. Property types and gatedness of the communities reflecting different community management were got through field trip observation, interviews with community managers and second data on websites. Measurements for these three variables then designed and determined which support the regression analysis for identification the factors influencing individual resilience (Table 4.10).

4.3.4.3 Determinants of Individual Resilience Identification (Individual Level)

Based on the previous research, we know the external attributes of community environment are potential influencing factors of the individual resilience. Therefore, to better identify the internal determinants of individual resilience on the individual level, Hierarchical Linear Model (HLM) is selected. HLM is a multilevel analysis method, which can disentangle the effects of independent variables operating at different spatial scales (Loo et al., 2017). Personal and individual-based local variables, which vary by subject, are input as Level 1 variables (the statistical unit is an individual). And the objective attributes of communities are considered as Level 2 variables (the statistical unit is a distinct local community).

Variables of Level 1 are from the questionnaires with the questions can get the variables demonstrating personal perceptions of urban flooding, the quality of physical environment and social environment. In this research, the dependent variable was the score of individual resilience performance, and the 31 independent variables were from the responses to the questionnaire. These factors are classified into three groups: three variables on the community scale (Table 4.9), the six control variables (Table 4.11) on the individual level, and the focused variables (Table 4.12) on the individual level. The focused variables on individual level are divided into six subgroups of explanatory variables: personal efficacy (PE), personal risk perception (PRP), social support perception (SSP), previous experiences of urban flooding (PEOF),

Table 4.9 Measurements of self-reported questions for individual resilience evaluation

Questions
(1) *Preparing for the potential* urban *flooding risk*
1. Do you do some preparation for heavy rain or urban flooding? No is scored 0; Yes is scored 10
2. Does your living community have the daily maintenance about the pipelines and pipe junctions? No is scored 0; not sure is scored 5; yes is scored 10
3. Does your living community have commissioners to repair the pipelines after they have been jammed?
No is scored 0; not sure is scored 5; yes is scored 10
(2) *Learnt to cope with and adapt to urban flooding*
4. Do you learn how to help yourself escape from urban flooding? No is scored 0; yes is scored 10
5. Do you take the initiative to study the knowledge of disaster prevention and risk aversion? No is scored 0; yes is scored 10
6. Does your living community provide education for improving your capacity for adapting to disasters? No is scored 0; not sure is scored 5; yes is scored 10
(3) *Past coping* strategies *during urban flooding*
7. Do you know where can you escape from the heavy rain or urban flooding when you are outside? No is scored 0; not sure is scored 5; yes is scored 10
8. Do you know which areas usually experience urban flooding around your living communities? No is scored 0; not sure is scored 5; yes is scored 10
9. Does your living community have some measurements to overcome the urban flooding? No is scored 0; not sure is scored 5; yes is scored 10
(4) *Habits to cope with and adapt to urban flooding*
10. Do you have the habit to check weather report? No is scored 0; yes is scored 10
11. Do you disseminate safe-haven news to your families and friends after you have known it? No is scored 0; not sure is scored 5; yes is scored 10
12. Do you know the early-warning signal of heavy rain? Don't know is scored 0; basically know is scored 3; know is scored 7; know well is scored 10
(5) *The creative and flexible* strategies *to adapt to urban flooding*
13. Do you follow the early-warning signal to arrange your travel flexibly? No is scored 0; yes is scored 10
14. Do you participate into the new strategies building for disaster prevention of your living community? No is scored 0; not sure is scored 5; yes is scored 10
15. Does your living community have some innovative strategies for adapting to disasters? No is scored 0; not sure is scored 5; yes is scored 10

Source Author

Table 4.10 Measurements of independent variable (community level)

Attributes of community environment	Measurement
Physical environment	
1. Drainage capacity (ordinal)	Scores: 0–10
Management	
1. Property types (ordinal)	Percentage of the commercial residential zones out of the total area of residential zones of a community
2. Gatedness of the communities (ordinal)	Percentage of the gated residential zones out of the total area of residential zones of a community

Source Author

Table 4.11 Controlling variables and their measurements

Variable	Measurement
Gender	Female (0), male (1)
Marital status	Single (0), marriage (1)
Age	>= 41 (0), <= 40 (1)
Income	>= 3000 (0), <3000 (1)
Occupation	No work (0), full time/part time job (1)
Education level	Equal or higher than professional education level (0), lower than professional education level (1)

Source Author

physical environment perceptions (PEP), and social cohesion perception (SCP). In order to carry out the HLM, the measurements of these variables are designed and determined (Tables 4.10, 4.11, and 4.12).

Table 4.12 Measurements of independent variables (individual level)

Personal perceived factors	Measurement
Personal efficacy (PE)	
1. Leadership (binary)	0 = no; 1 = yes
2. Ability to lead others to escape from hazard during urban flooding (binary)	0 = no; 1 = yes
3. Public participation demands on related urban planning and management (binary)	0 = no; 1 = yes
Personal risk perception (PRP)	
4. Feelings about urban flooding (ordinal)	1 = no dangerous 2 = not very dangerous 3 = dangerous 4 = very dangerous
5. Sensitive to urban flooding (binary)	0 = no; 1 = yes
6. Concerns about children during urban flooding (binary)	0 = no; 1 = yes
7. Concerns about the elderly during urban flooding (binary)	0 = no; 1 = yes
Social support perception (SSP)	
8. Help and support from the government (binary)	0 = do not have; 1 = have
9. Help and support from their workplaces (binary)	0 = do not have; 1 = have
Previous experiences of urban flooding (PEOF)	
10. Number of days worked per week (ordinal)	1 = > 5 days 2 = 5 days 3 = < 5 days
11. The primary means of commuting (nominal)	1 = public transportation 1 = bike or walk 1 = private cars 1 = others
12. Experienced urban flooding (binary)	0 = no; 1 = yes
13. Previous losses during urban flooding (ordinal)	1 = none 2 = not too much 3 = general 4 = much 5 = very much
Physical environment perceptions (PEP)	
14. Amount of construction projects in progress around communities (ordinal)	1 = quite a lot 2 = general 3 = not too many
15. Areas of ecological lands around communities (binary)	0 = not too many; 1 = a lot

(continued)

Table 4.12 (continued)

Personal perceived factors	Measurement
16. Living environment of communities (ordinal)	1 = very good 2 = good 3 = general 4 = bad 5 = very bad
17. Waste management created by natural hazards of communities (ordinal)	1 = very clean 2 = clean 3 = general 4 = not clean 5 = very unclean
Social cohesion perception (SCP)	
18. Mutual trust during times of difficulty (binary)	0 = do not have; 1 = have
19. Mutual help during times of difficulty (binary)	0 = do not have; 1 = have
20. Perception of life in your residential area (ordinal)	1 = very bad 2 = bad 3 = general 4 = good 5 = very good
21. *Hukou* status	0 = do not have Shenzhen *hukou*; 1 = have
22. Housing ownership	0 = do not have; 1 = have

Source Author

4.4 Conclusion

This chapter proposes an analytical framework to test three research hypotheses and answer three groups of research questions: (1) What was the current performance of individual resilience of residents vulnerable to urban flooding? And did the individual resilience vary among different communities? (2) What were the determinants (People-Disaster-Place-Society) of individual resilience? (3) Whether the strategic measure of Sponge City Initiative could help nurture individual resilience? And what were the place-responsive measures for individual resilience nurturing?

In order to test research hypotheses and answer the research questions, a detailed research approach is provided. More specifically, in the first stage of analysis, individual resilience is evaluated based on the self-reported coping and adaptive strategies adopted by individuals and their communities. Further, the variations of individual resilience in different communities are explored and examined. In the second stage, the determinants of individual resilience are identified. Unlike similar research, the potential determinants identified reflect personal perceptions (People) of urban flooding (Disaster), quality of physical environment (Place) and social environment (Society). In the third stage, whether Sponge City Initiative can help nurture individual resilience is explored based on the previous analyses on the interactions among people, disaster, place and society. This could provide empirical support for formulating policies and strategies to improve individual resilience.

This chapter also describes the research methods adopted, including data sources, data collection methods, case selection, and data analysis methods. Both quantitative and qualitative methods were used in the analysis. Data sources included the spatial data of DEM, meteorological data, land use and statutory plans, survey data, site observations, semi-structured interviews, and desktop research. The dependent and independent variables for statistical analysis were then got through questionnaire, or model simulation with the help of ArcGIS, SPSS and SWMM in this research. Determinants of individual resilience are then explored with the help of HLM. Interview is the main method to explore the effectiveness of Sponge City Initiative on individual resilience nurturing.

Chapter 5
Individual Resilience to Urban Flooding of Gongming

5.1 Introduction

Urban flooding not only threatens property, but also human health and their well-being. Previous research and practices have emphasized the improvement of infrastructure to prevent urban flooding. To say nothing about the difficulty of such improvements, the increasingly unsure circumstances associated with climate change make certain that no city can be made impregnable. In the past 40 years, especially in western countries, resilience has been expressed as the capacities, strategies, and methods to cope with and adapt to unexpected disturbances by people who are vulnerable to disasters like urban flooding. Therefore, it is necessary to evaluate resilience of residents, which can reflect the capacities of residents to cope with and adapt to urban flooding.

Gongming is located in Guangming New District of Shenzhen. It is in the center of the new district of Shenzhen, which suffers from urban flooding every year. In addition, more than 80% of the dwellings are small property housing, accommodating mainly migrant workers. All these situations make it attractive to explore the characteristics of individual resilience for this type of social groups.

As explained in the framework, individual resilience is the first step to realize the aim of this research. In this chapter, the major findings of the analysis are presented to answer the first group of research questions: (1) What was the current performance of individual resilience of residents vulnerable to urban flooding? And did the individual resilience vary among different communities? Individual resilience, the multi-dimensional capacity for coping and adaptation, is evaluated based on the judgment on whether the human behavior is resilience-oriented. Strategies of coping and adaptation adopted by individuals and their communities are identified based on the interviews with residents, governments and scholars. Then individual resilience can be evaluated based on the selection of strategies by local residents through questionnaire. Further, the performance of individual resilience of the population is evaluated based on the current resilience evaluation of the sample. Finally,

J. Song, *Individual Resilience to Urban Flooding and the Implications for Urban Management*, https://doi.org/10.1007/978-981-97-0546-7_5

average individual resilience performance level of residents of different communities is assigned to the each community. The spatial variation of individual resilience among communities then can be identified and explored.

5.2 Descriptive Analyses on Adopted Coping and Adaptation Strategies

The identification of human behavior, i.e. coping and adaptation strategies against urban flooding, is achieved by consulting scholars and local residents, and referencing similar researches and documents of the government. These strategies are just part of all the possible strategies adopted by individuals, but cover the most important 15 coping and adaptation strategies. Based on the theory of adaptive cycle and Fig. 2. 2, strategic capacity is positively related to, everyday capacity so do adaptation capacity and coping capacity. Therefore, in this research, coping is not distinguished from adaptation. All these strategies together are marked according to the scoring methods and standards introduced in Chap. 4. To understand the differences among individual strategies in Gongming, this study first ranked the frequencies of the 15 strategies cited by the surveyed residents and calculated the percentages of the surveyed residents citing those strategies (Table 5.1).

Here according to the figure reflecting the intensity of application of the strategies, these strategies can be divided into three groups: most adopted (70–100%, including 70%), generally adopted (30–70%, including 30%) and least adopted (0–30%, including 0%).

The most adopted group of strategies consists of "disseminate safe-haven news to your families and friends after you have known it," "do some preparation for heavy rain or urban flooding," "have the habit to check weather report" and "follow the early-warning signal to strictly arrange your travel." With the survey data, we can conclude that with the popularity of smartphones, information sharing and communication has become much easier. Most residents expressed willingness to heed the early-warning signals and prepare for urban flooding. It seems that the residents in Gongming have a certain degree of disaster awareness.

The least adopted group of strategies consists of "know the early-warning signal of heavy rain," "know which areas usually experience urban flooding around my living communities," "my living community has some innovative strategies for adapting to disasters," "my living community provides education for improving your capacity for adapting to disasters," "my living community has the daily maintenance about the pipelines and pipe junctions" and "my living community has some measurements to overcome the urban flooding." These strategies are reported by no more than 30% of the residents. We can therefore conclude that although 84.17% of the residents show a willingness to follow the early-warning signals, only 7.94% of the residents claim to know how to recognize those signals. Only 22% of the surveyed residents reported that they know which areas of their neighborhood are likely to be flooded.

Table 5.1 Self-reporting coping and adaptation strategies adopted by residents in Gongming

Strategies	Percentage (%)[a]
(1) *Preparing for the potential urban flooding risk*	
1. Do you do some preparation for heavy rain or urban flooding?	87.18
2. Does your living community have the daily maintenance about the pipelines and pipe junctions?	19.10
3. Does your living community have commissioners to repair the pipelines after they have been jammed?	42.43
(2) *Learnt to cope with and adapt to urban flooding*	
4. Do you learn how to help yourself escape from urban flooding?	35.18
5. Do you take the initiative to study the knowledge of disaster prevention and risk aversion?	68.21
6. Does your living community provide education for improving your capacity for adapting to disasters?	15.14
(3) *Past coping strategies during urban flooding*	
7. Do you know where can you escape from the heavy rain or urban flooding when you are outside?	38.74
8. Do you know which areas usually experience urban flooding around your living communities?	22.10
9. Does your living community have some measurements to overcome the urban flooding?	18.96
(4) *Habits to cope with and adapt to urban flooding*	
10. Do you have the habit to check weather report?	73.53
11. Do you disseminate safe-haven news to your families and friends after you have known it?	90.86
12. Do you know the early-warning signal of heavy rain?	7.94
(5) *The creative and flexible strategies to adapt to urban flooding*	
13. Do you follow the early-warning signal to arrange your travel flexibly?	84.17
14. Do you participate into the new strategies building for disaster prevention of your living community?	38.20
15. Does your living community have some innovative strategies for adapting to disasters?	4.91

Notes [a]The percentage of selected individuals adopted that coping or adaptation strategy

From one aspect, these choices suggest the ability of residents to cope with and adapt to urban flooding is not high even they are willing to do so. From another apsect, the self-reported strategies reflect that most residents do not know or do not think their living communities have adopted strategies to cope with or adapt to urban fooding. It suggests that on the one hand the participation of residents into their living communities are seldom. On the other hand, living-community-level urban flood management, e.g. maintaining the sewage system, preventing floodwater from falling into underground spaces (e.g. garages) in time, etc. may be partly ignored in our research area.

35.18–68.21% of the respondents reported the following strategies: "have learnt how to help myself escape from disasters," "know where I can escape from the heavy rain or urban flooding when I am outside," "take the initiative to study the knowledge of disaster prevention and risk aversion," "I am willing to participate into the new strategies building for disaster prevention of my living community" and "my living community has commissioners to repair the pipelines after they have been jammed." Only 20% of the selected residents know which areas were flooded and 30% know how to help themselves when caught in a flood. Two-thirds of the residents showed a willingness to learn about disaster and risk prevention. These further verify that residents show a willingness to learn the strategies to cope with and adapt to urban flooding, but the present controllign ability of residents on coping with and adaptation to urban flooding is low.

In summary, the self-reported strategies adopted by residents of Gongming show that they have the willingness to cope with and adapt to urban flooding, but lack the ability to do so. Since the human beings can improve their ability through learning within social groups around them, it suggests that there should be some common attributes among these residents. Besides, it also suggests the living communities seems play a role on this type of human behavior. This is the main focus of discussion in the following part of this chapter.

5.3 Individual Resilience Evaluation

As reviewed in Chap. 2, the selection of strategies reflects the capacity of individuals to withstand heavy rain and urban flooding. The choice of the strategies is not only influenced by the individuals themselves, but also the coorperators, such as family members, neighbors or help from their living communities, which can benefit, supplement and facilitate individual behavior towards the desired level of resilience. After identifying the self-reported coping and adaptation strategies, based on the selection and application of these strategies by individuals, the scores for their individual resilience are calculated as described in Chap. 4. The descriptive statistics of the individual resilience of 733 residents in Gongming are shown in Table 5.2.

According to Table 5.2, the range of individual resilience scores in the research area was 9.80, with the lowest score = 0.20 and the highest score = 10.00. The higher the score, the higher individual resilience performance level. In this research, the original scores were dichotomized: low (0–3, including 3) versus high (7–10, including 7), with one broad intermediate level (3–7). The mean score of individual resilience for the research area was 4.62 (intermediate level). The median was 4.67, reflecting that most residents' individual resilience is at intermediate level. Compared to the normal distribution, the distribution of the individual resilience values shows a sharp peak and flat tails through the kurtosis value. In addition, the skewness value together with the histogram (Fig. 5.1) demonstrates several extremely high values

Table 5.2 Descriptive statistics of individual resilience in Gongming

Descriptive	Statistic	Individual resilience (n = 733)
Range	Statistic	9.80
Minimum	Statistic	0.20
Maximum	Statistic	10.00
Mean	Statistic	4.62
Median	Statistic	4.67
Std. Deviation	Statistic	1.67
Variance	Statistic	2.79
Skewness	Statistic	0.27
	Std. error	0.09
Kurtosis	Statistic	0.08
	Std. error	0.18

Source Author

on the right. This means that several residents in the study area have some basic strategies to overcome urban floods quite well, but there are still a large number of residents with intermediate, low, or even extremely low level of individual resilience performance.

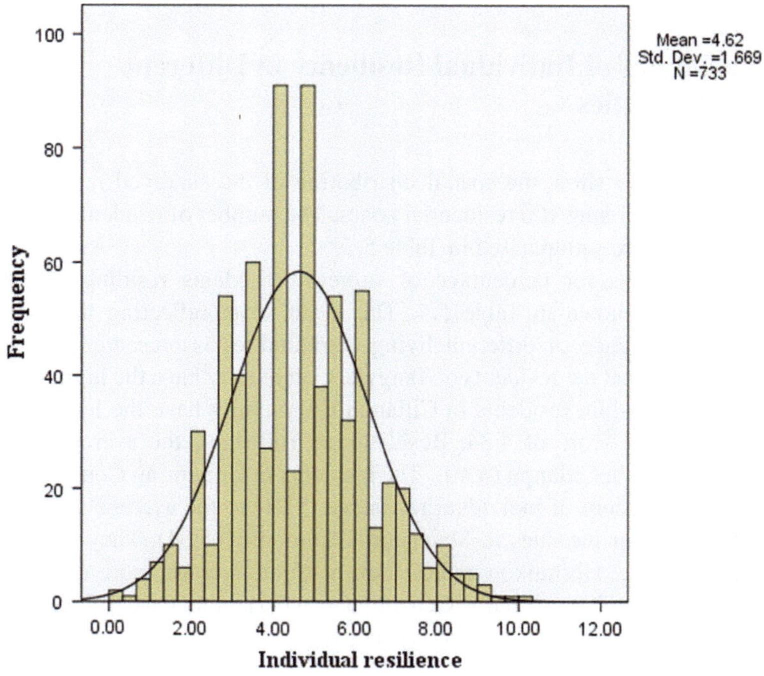

Fig. 5.1 Histogram of individual resilience. *Source* Author

In summary, the individual resilience of the selected research area is at, or even below, the intermediate level. It means that most of the Gongming residents have not gained some basic strategies to cope with, or have the willingness or ability to adapt to urban flooding. Although there are several residents with extremely high individual resilience (even attaining the full mark of 10.00), according to the "Buckets Effects", the individuals with low resilience should be focused on. Their attributes can help to explain the low individual resilience as well, which is also the focus in the following chapter. In sum, there is still room for individual resilience to grow, and this it is necessary for Gongming to enhance the level of individual resilience performance.

In summary, although people seem to believe they have the ability to withstand urban flooding, they lack the knowledge and resources to do so. Meanwhile, cooperators within the residents seem do not play the due role to support individual resilience. Most residents report that they do not know or lack the strategies of their living zones to cope with and adapt to urban flooding. The average level of individual resilience performance of residents of Gongming falls into the intermediate level, below the passing score of 6.00/10.00. Moreover, the statistics show that there are people in Gongming with extremely high capacities to cope with or adapt to urban flooding. Nonetheless, most residents in the study area lack high individual resilience, posing potential danger to themselves and to others. Therefore, it is necessary for us to identify the characteristics of these residents. The factors limiting their individual resilience will be identified in the following chapter.

5.4 Variations of Individual Resilience in Different Communities

Figures 5.2 and 5.3 show the spatial distribution of the surveyed residents in 11 communities comprising 100 residential zones. The number of residential zones and their distribution are summarized in Table 5.3.

Statistical results for percentage of surveyed residents residing in different communities are shown in Table 5.4. The mean score reflecting the individual resilience performance of different living communities is presented in Fig. 5.4. Figure 5.4 shows that the residents of Tangwei Community have the highest average resilience (5.63), while residents in Citianpu Community have the lowest average resilience with the score of 4.34. Residents in Yulv have the average resilience similar to the ones in Citianpu (4.40). The residents in Gongming Community have a similar average score of individual resilience (5.01) to the average score of individual resilience for the ones in Shangcun Community (5.04). They are followed Jiangshi, Hongxing, Heshuikou and Xiacun, with an average score of 4.98, 4.88, 4.87 and 4.82 respectively. Finally, Genzhuyuan and Tianliao have a similar average individual resilience of 4.62 and 4.69 respectively.

Fig. 5.2 Spatial distribution of residential zones of selected residents. *Source* Author

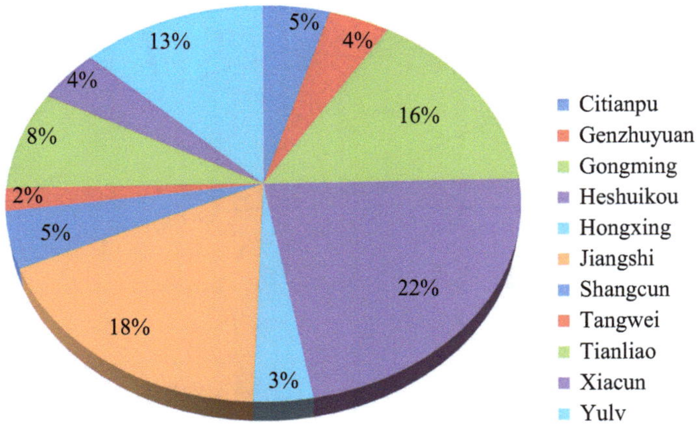

Fig. 5.3 The percentage of surveyed residents residing in different communities. *Source* Author

Table 5.3 Residential zones in different communities of the study area

Community name	No. of residential zones	No. of surveyed residents	Average surveyed residents per residential zone
Citianpu	2	34	17
Genzhuyuan	2	31	16
Gongming	17	116	7
Heshuikou	21	165	8
Hongxing	3	25	8
Jiangshi	20	129	6
Shangcun	7	34	5
Tangwei	3	14	5
Tianliao	10	59	6
Xiacun	4	32	8
Yulv	11	94	9
Total	100	733	7

Source Author

There exists a pattern of spatial distribution for these communities' level of resilience as well (Fig. 5.5). Residents of communities in the middle part of Gongming (Tangwei and Jiangshi) have high individual resilience, followed by those in the northern part (Shangcun, Xiacun, Gongming and Heshuikou). The residents of the northwestern and southern parts of Gongming, are having lower individual resilience. These communities include Citianpu, Genzhuyuan, Yulv and Tianliao. In essence, neighboring communities usually have similar geographical situations, built environment and socio-demographic conditions. Therefore, we can conclude that such

Table 5.4 Statistical results for different communities

Community	N	Mean	Std. deviation	Variance	Minimum	Maximum
Citianpu	34	4.34	1.51	2.27	1.67	7.45
Genzhuyuan	31	4.62	1.90	3.60	0.53	8.53
Gongming	116	5.01	1.59	2.54	1.53	8.53
Heshuikou	165	4.87	1.56	2.43	0.67	10.00
Hongxing	25	4.88	1.63	2.66	1.87	8.33
Jiangshi	129	4.98	1.43	2.03	0.20	9.78
Shangcun	34	5.04	1.66	2.75	2.53	9.20
Tangwei	14	5.63	1.63	2.66	0.78	7.20
Tianliao	59	4.69	1.66	2.75	2.11	9.33
Xiacun	32	4.82	1.44	2.08	1.67	7.76
Yulv	94	4.40	1.59	2.52	1.00	9.00
Total	733					

Source Author

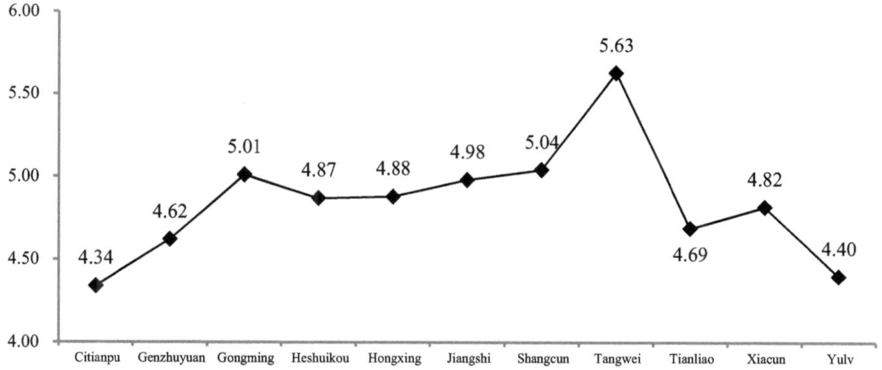

Fig. 5.4 Individual resilience of different communities. *Source* Author

common attributes of communities may distinguish the residents with low resilience from the ones with high resilience.

5.5 Discussion and Conclusion

This study presents a quantitative method to assess the current performance of individual resilience to urban flooding with the case of Gongming, and includes coping capacity and adaptation capacity based on the identification of their resilience-oriented behavior. Following the assessment of individual resilience, this chapter

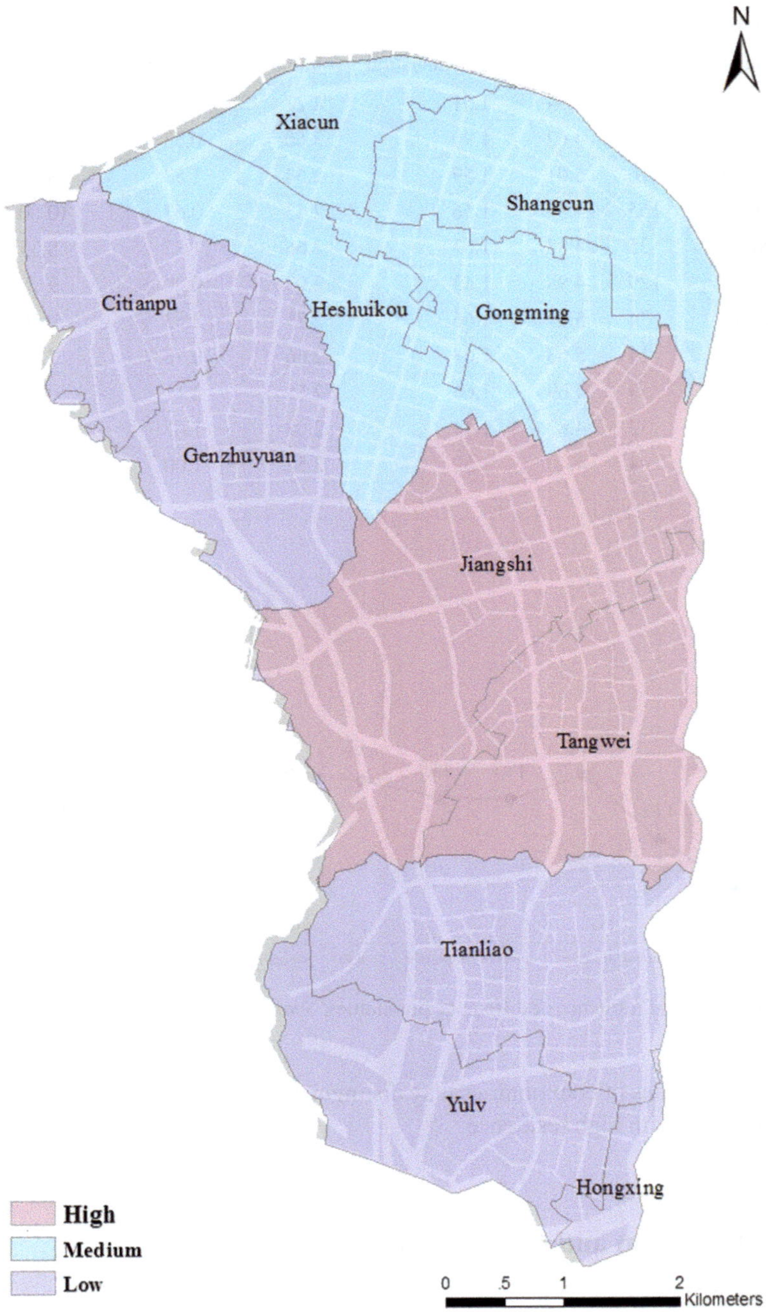

Fig. 5.5 Spatial distribution of communities with different individual resilience scores. *Source* Author

has answered the first group of research questions: (1) What was the current performance of individual resilience of residents vulnerable to urban flooding? And did the individual resilience vary among different communities? This part of research has tested the first research hypothesis that there are significant differences in individual resilience to urban flooding for residents in different communities.

Results show that most individuals of Gongming are willing to behave in a resilience-oriented manner but lack the ability to do so. Overall, the individual resilience performance level of the study area is not high, falling only in the low intermediate level and far from the passing score of 6 out of 10. Based on the results of performance of individual resilience, we find that individual resilience varies across different communities and shows spatial variability. The study area comprises 11 communities and the results suggest that the communities located in the middle part of Gongming show the highest individual resilience, followed by residents of communities located in the northern part of Gongming. The residents of communities in the northwest and southern part of Gongming are more likely to have low individual resilience.

More conclusions can be drawn based on the results:

Given the high vulnerability to urban flooding for residents of Gongming, the relatively low individual resilience may not be able to withstand the potential natural disasters. Therefore, it is necessary to improve the individual resilience of people living in Gongming. In fact, the low individual resilience of Gongming itself means there is still room for their individual resilience to improve.

One social group has high willingness but limited ability to adapt to urban flooding. This finding initiates thereby that the lack of education may be a limiting factor of residents' individual resilience performance. More deeply, mission-learning paths may be one of the reasons leading to their low resilience performance level. If so, individual resilience is relatively easy to be improved through external interventions, such as trainings for resilience nurturing provided by formal and informal organizations. Another social group has high ability to learn to adapt to urban flooding but is unwilling to do so. Although this group of people is not large in number, compared to people who lack the ability to cope with and adapt to urban flooding, it seems difficult for them to improve their individual resilience. We must identify the reasons for their low willingness to behave in a resilience-oriented manner. Therefore, the characteristics of this group of the population deserve more attention.

The evaluation of individual resilience is the foundation of this study. Based on the quantitative evaluation of resilience for the study area, we can draw the conclusions mentioned above. Besides, finding proves that there are spatial disparities among 11 communities of Gongming. This finding suggests that common attributes of people's communities distinguish the residents with high individual resilience from the ones with low individual resilience performance level. These conclusions provide the clues and contexts for the following research, and the factors limiting individual resilience performance in Gongming will be presented in the next chapter. Based on these limiting factors, the efficiency and effects of urban management to change the current level of resilience will be examined and possible pathways to improvement on the community scale be described in Chap. 7.

Chapter 6
Factors Influencing Individual Resilience to Urban Flooding: People-Disaster-Place-Society

6.1 Introduction

Based on the results reported in Chap. 5, the spatial distribution of individual resilience in different communities suggests the attributes of communities should be the potential determinants of individual resilience. Except the attributes on the community level, the second hypothesis of this research: individual perceptions of urban flooding, living place and living society can influence individual resilience. In order to test this research hypothesis, in this chapter I identify the factors that encourage and inhibit resilience, and to achieve this, I need to answer the second research questions: "What were the determinants (People-Disaster-Place-Society) of individual resilience?"

Quantitative and qualitative methods, including statistical analyses, multiple regressions based on questionnaires and field studies, are combined to explore the factors that influenced and limited individual resilience. In this study, the characteristics that distinguish different communities are identified in terms of the drainage capacities, gatedness and the property types of communities. The drainage capacities of sewage systems are calculated by SWMM as described in Appendix C, and the property types and gatedness of the community are identified through the observation and interview during the field trip with supplementary data on websites. Data on socioeconomic characteristics, as the control variables, as well as the potential factors limiting individual resilience are obtained through questionnaires. Then, the factors that influenced and limited individual resilience are identified on the individual level, also considering the different attributes of community level with Hierarchical Linear Model (HLM).

J. Song, *Individual Resilience to Urban Flooding and the Implications for Urban Management*, https://doi.org/10.1007/978-981-97-0546-7_6

6.2 Factors that Influenced Individual Resilience

6.2.1 Socio-demographic Characteristics of Gongming Residents: The Control Variables

Chapter 2 explains that some socio-demographic characteristics are significant individual determinants of resilience performance, so their effects should be controlled for to identify the influences of the key variables selected for this study. For the analysis, six socio-demographic factors were selected as control variables: gender, marital status, age, income, occupation and education level.

Control variables were analyzed based on their statistical descriptions. The sample consisted of 443 males and 290 females (male to female ratio of 1.53). In the sample, 444 residents were married (60.6%) and 289 residents were unmarried. There were four age groups—about 2.7% being 16 to 18, 82.7% being 18 to 40, and 14.6% being 40 or above. One third of the surveyed residents have an income of no more than 3000 CNY/month, and half (53.5%) have 3000–7000 CNY/month. It means most of the surveyed residents (85.7%) have a monthly income of less than CNY 7000. More than one-half (65.7%) was working, including 5.9% as part-time workers. A quarter (25.4%) of the surveyed residents were educated up to middle school, and 49.0% of them completed high school. Only 25.6% of the surveyed residents hold at least a professional degree. In general, the population is not highly educated (Table 6.1).

The male to female ratio of the sample was higher than the average of Guangming New District in 2010 (1.23).[1] In terms of age, 82.7% of the sample was between 18 and 40 years old, higher than Gongming Sub-district's average (about 70%). Because the Gongming study site comprised only 11 communities of the Gongming Sub-district (total 19 communities, before 2016),[2] which constitute the economic center of Guangming New District with high population density, and this should explain why there were more young people and more males (main labor force) than the statistical data of Guangming New District. Of these residents, 85% of them have income lower than 7000 CNY/month (the average wage in Shenzhen is 6753 CNY/month).[3] 76% have attained high school education, which is lower than the average level of Shenzhen in 2010 according to the sixth census data (around 82.82%).[4] Considering that the research area attracts quite a lot of peasant workers instead of talents like other places of Shenzhen, the survey data seems reasonable.

Before performing the regression analyses, all of the categorical control variables were transformed into dummy variables (Table 6.2).

---

[1] Source: The sixth census of Shenzhen, 2010, retrieved from http://www.sztj.gov.cn/xxgk/tjsj/pcgb/201606/t20160614_3697014.htm.

[2] https://goo.gl/SLNeFn.

[3] http://www.szhrss.gov.cn/xwdt/201606/t20160627_3724221.htm.

[4] http://www.sztj.gov.cn/xxgk/tjsj/pcgb/201105/t20110512_2061597.htm.

Table 6.1 Descriptive statistics of the control variables

		n	Valid percent
Gender	Male	443	60.4
	Female	290	39.6
	Total	733	100.0
Marital status	Married	444	60.6
	Single	289	39.4
	Total	733	100.0
Age	16–18	20	2.7
	18–40	606	82.7
	>= 40	107	14.6
	Total	733	100.0
Income	<= 3000 CYN/month	236	32.2
	3000–7000 CYN/month	392	53.5
	7000–15,000 CYN/month	98	13.4
	>= 15,000 CYN/month	7	0.9
	Total	733	100.0
Occupation	At work	438	59.8
	Part-time	43	5.9
	No work	252	34.3
	Total	733	100.0
Education level	Equal or lower than middle school	186	25.4
	High school	359	49.0
	University	186	25.4
	Higher than university	2	0.2
	Total	733	100.0

Source Author

Table 6.2 Control (dummy) variables in the overall sample (N = 733)

Variable	Measurement	No. of measurement = 1	Percentage (%)
Gender	Female (0), male (1)	443	60.44
Marital status	Single (0), marriage (1)	444	60.57
Age	>= 41 (0), <= 40 (1)	626	85.40
Income	>= 3000 (0), < 3000 (1)	236	32.20
Occupation	No work (0), fulltime/part-time job (1)	481	65.62
Education level	Equal or higher than professional education level (0) Lower than professional education level (1)	545	74.35

Source Author

Table 6.3 Drainage capacity of each community

Community	Drainage capacity	Level
Citianpu	7.21	High
Genzhuyuan	0.32	Low
Gongming	5.50	Medium
Heshuikou	4.68	Medium
Hongxing	4.80	Medium
Jiangshi	3.28	Medium
Shangcun	2.65	Low
Tangwei	3.04	Low
Tianliao	6.69	Medium
Xiacun	3.44	Medium
Yulv	8.56	High
Average	4.56	Medium

Source Author

6.2.2 Factors Reflecting Community Attributes

In this study, external environment of living community are focused on to show the attributes of communities for different residents. Drainage capacity, gatedness and property types of the community are selected as the potential factors influencing individual resilience of people residing in the community.

6.2.2.1 Physical Environment

Drainage capacity is selected to reflect the physical environment of the research area. Appendix C details the evaluation process of the drainage capacity for this study area. Average drainage capacity of residents of different communities is assigned to the each community, and the drainage capacity for each community is shown in Table 6.3.

6.2.2.2 Social Environment

The level of social environment can be reflected through gatedness and property types of communities. The information and data of the 733 residents' communities (11 communities) are obtained through site observation and interviews, supplemented by the secondhand data and information on websites. The percentage of gated residential zones out of the total area of residential zones of a community is assigned to the each community, which can reflect the gatedness level of each community (Table 6.4). By the same token, the percentage of commercial residential zones out of the total area

Table 6.4 Statistical data for gatedness of each community

Community	Percentage (%)	Level
Citianpu	100.00	Low
Genzhuyuan	100.00	High
Gongming	65.79	Medium
Heshuikou	58.28	Medium
Hongxing	88.96	Low
Jiangshi	66.67	Medium
Shangcun	21.38	Medium
Tangwei	36.22	Medium
Tianliao	13.98	Low
Xiacun	2.88	Low
Yulv	62.18	Medium
Average	56.03	Medium

Source Author

of residential zones of a community is assigned to the each community, which can reflect the main type of property for each community (Table 6.5).

Statistical analyses show that 43.97% of the surveyed residents live in gated communities and the other 56.03% do not, showing an almost half-and-half distribution for these two types of property management. Besides, 83.15% of the surveyed residents' residential zones are non-commercial property; 16.85% of them are commercial property. The ones living in non-commercial property outnumber those living in commercial property.

Table 6.5 Statistical data for property types of each community

Community	Percentage (%)	Level
Citianpu	0.00	Low
Genzhuyuan	0.00	Low
Gongming	56.28	Medium
Heshuikou	41.32	Medium
Hongxing	0.00	Low
Jiangshi	19.98	Low
Shangcun	18.22	Low
Tangwei	29.34	Low
Tianliao	13.98	Low
Xiacun	2.88	Low
Yulv	3.36	Low
Average	16.85	Low

Source Author

6.2.2.3 Environment of Communities

Table 6.6 shows the types of communities in terms of social environment and physical environment on the communal level.

Finally, seven types of communities are formed based on the selected factors, i.e. HML, HLL, MMM, MML, MLL, LHL, and LML. Yulv belongs to the type of HML; Citianpu belongs to type of HLL; Gongming and Heshuikou belong to type of MMM; Jiangshi belongs to MML; Hongxing, Tianliao and Xiacun belong to type of MLL; Genzhuyuan belongs to LHL; Tangwei and Shangcun belong to type of LML. Overall, Gongming belongs to the type of MML.

6.2.3 Factors Reflecting Personal Perceptions

After identifying the control variables and factors reflecting community attributes, our attention turned to the effects of the individual characteristics on the performance of individual resilience. The hypothesis tested by regression analysis is that not only the residents' socio-ecological characteristics influenced the current performance of individual resilience, but also the personal perceptions of urban flooding, physical environment and community management.

Based on previous research, in this research, 22 variables reflecting personal perceptions, comprising a total of six explanatory factors (PE, PRP, SSP, PEOF, PEP, and SCP), were derived from the questionnaire data. The descriptive statistics of these variables and the measurements of the variables also shown in Table 6.7, which are used to test the associations with individual resilience performance.

Table 6.6 Types of communities

Community	Drainage capacity	Gatedness	Property types
Citianpu	High	Low	Low
Genzhuyuan	Low	High	Low
Gongming	Medium	Medium	Medium
Heshuikou	Medium	Medium	Medium
Hongxing	Medium	Low	Low
Jiangshi	Medium	Medium	Low
Shangcun	Low	Medium	Low
Tangwei	Low	Medium	Low
Tianliao	Medium	Low	Low
Xiacun	Medium	Low	Low
Yulv	High	Medium	Low
Average	Medium	Medium	Low

Source Author

Table 6.7 Descriptive statistics of independent variables (individual level) in the overall sample

Personal perceived factors	Measurement	N (733)	Percentage
Personal efficacy (PE)			
1. Leadership (binary)	0 = no; 1 = yes	466	63.57
2. Ability to lead others to escape from hazard during urban flooding (binary)	0 = no; 1 = yes	373	50.89
3. Public participation demands on related urban planning and management (binary)	0 = no; 1 = yes	405	55.25
Personal risk perception (PRP)			
4. Feelings about urban flooding (ordinal)	1 = no dangerous 2 = not very dangerous 3 = dangerous 4 = very dangerous	135 449 108 41	18.42 61.26 14.73 5.59
5. Sensitive to urban flooding (binary)	0 = no; 1 = yes	426	58.12
6. Concerns about children during urban flooding (binary)	0 = no; 1 = yes	446	60.85
7. Concerns about the elderly during urban flooding (binary)	0 = no; 1 = yes	438	59.75
Social support perception (SSP)			
8. Help and support from the government (binary)	0 = do not have; 1 = have	27	3.68
9. Help and support from their workplaces (binary)	0 = do not have; 1 = have	150	20.46
Previous experiences of urban flooding (PEOF)			
10. Number of days worked per week (ordinal)	1 = > 5 days 2 = 5 days 3 = < 5 days	304 218 211	41.47 29.74 28.79
11. The primary means of commuting (nominal)	1 = public transportation 1 = bike or walk 1 = private cars 1 = others	283 297 114 39	38.61 40.52 15.55 5.32
12. Experienced urban flooding (binary)	0 = no; 1 = yes	341	46.52
13. Previous losses during urban flooding (ordinal)	1 = none 2 = not too much 3 = general 4 = much 5 = very much	540 109 30 26 28	73.67 14.87 4.09 3.55 3.82
Physical environment perceptions (PEP)			
14. Amount of construction projects in progress around communities (ordinal)	1 = quite a lot 2 = general 3 = not too many	200 317 216	27.29 43.25 29.47

(continued)

Table 6.7 (continued)

Personal perceived factors	Measurement	N (733)	Percentage
15. Areas of ecological lands around communities (binary)	0 = not too many; 1 = a lot	293	39.97
16. Living environment of communities (ordinal)	1 = very good	27	3.68
	2 = good	253	34.52
	3 = general	390	53.21
	4 = bad	55	7.50
	5 = very bad	8	1.09
17. Waste management created by natural hazards of communities (ordinal)	1 = very clean	36	4.91
	2 = clean	341	46.52
	3 = general	299	40.79
	4 = not clean	50	6.82
	5 = very unclean	7	0.95
Social cohesion perception (SCP)			
18. Mutual trust during times of difficulty (binary)	0 = do not have; 1 = have	694	94.68
19. Mutual help during times of difficulty (binary)	0 = do not have; 1 = have	704	96.04
20. Perception of life in your residential area (ordinal)	1 = very bad	2	0.27
	2 = bad	17	2.32
	3 = general	397	54.16
	4 = good	272	37.11
	5 = very good	45	6.14
21. *Hukou* status	0 = do not have Shenzhen *hukou*; 1 = have	82	11.19
22. Housing ownership	0 = do not have; 1 = have	177	24.15

Source Author

6.3 Associations Between Individual Resilience and Focusing Factors

6.3.1 Personal Perceptions and Individual Resilience

HLM was chosen to identify and explore the associations between personal perceptions and individual resilience (the dependent variable) after considering the influences of different types of communities. Except the control variables, the other variables were sorted into six explanatory factors: (1) PE, (2) PRP, (3) SSP, (4) PEOF, (5) PEP, and (6) SCP. The regression models are shown in Table 6.8.

Table 6.8 shows the explanatory power of the control variables. Among these six control variables, only gender is a significant variable to the changes of individual resilience performance. When the measurements of the control variables in Table 6.7 and the coefficients from regression (seen in Table 6.8) analysis are combined, we can conclude that males are more likely to have high individual resilience performance

Table 6.8 The estimations of the determinants of individual resilience

Independent variables	Coefficient	Standard Error	T-ratio	Approx. d.f	*P*-value
Fixed-effects					
INTRCPT	4.838	0.562	8.603	7	0.000***
Drainage capacity	– 0.019	0.016	1.215	7	0.271
[a]Gatedness of the community	0.288	0.131	2.200	7	0.069*
Property types	0.265	0.408	0.422	7	0.160
Control variables					
Gender	0.355	0.114	3.125	699	0.002***
Marital status	0.165	0.122	1.348	699	0.178
Age	0.123	0.166	0.740	699	0.460
Income	0.007	0.150	0.045	699	0.964
Occupation	– 0.040	0.158	– 0.251	699	0.802
Education level	– 0.037	0.082	– 0.447	699	0.655
Personal efficacy (PE)					
Leadership	0.533	0.115	4.652	699	0.000***
Ability to lead others to escape from hazard during urban flooding	– 0.049	0.100	– 0.492	699	0.622
Public participation demands on related urban planning and management	0.250	0.108	2.314	699	0.021**
Personal risk perception (PRP)					
Feelings about urban flooding	– 0.060	0.079	– 0.751	699	0.453
Sensitive to urban flooding	0.053	0.118	0.452	699	0.651
Concerns about children during urban flooding	0.274	0.139	1.979	699	0.048**
Concerns about the elderly during urban flooding	0.044	0.140	0.313	699	0.754
Social support perception (SSP)					
Help and support from the government	0.522	0.284	1.833	699	0.067*
Help and support from their workplaces	0.644	0.135	4.766	699	0.000***
Previous experiences of urban flooding (PEOF)					
Number of days worked per week	0.079	0.065	1.217	699	0.224
The primary means of commuting is buses	– 0.029	0.244	– 0.118	699	0.906

(continued)

Table 6.8 (continued)

Independent variables	Coefficient	Standard Error	T-ratio	Approx. d.f	*P*-value
The primary means of commuting is walk or bike	− 0.164	0.244	− 0.670	699	0.503
The primary means of commuting is private cars	− 0.038	0.274	− 0.138	699	0.890
Experienced urban flooding	0.255	0.114	2.242	699	0.025**
Previous losses during urban flooding	0.172	0.055	3.139	699	0.002***
Physical environment perceptions (PEP)					
Amount of construction projects in progress around communities	− 0.117	0.075	− 1.569	699	0.117
Areas of ecological lands around communities	0.000	0.113	0.003	699	0.997
Living environment of communities	− 0.453	0.095	− 4.780	699	0.000***
Waste management created by natural hazards of communities	− 0.163	0.090	− 1.810	699	0.070*
Social cohesion perception (SCP)					
Mutual trust during times of difficulty	0.674	0.199	3.384	699	0.001***
Mutual help during times of difficulty	0.205	0.258	0.797	699	0.426
Perception of life in your residential area	0.147	0.077	− 1.904	699	0.057*
Shenzhen *hukou*	− 0.003	0.171	− 0.016	699	0.987
Housing ownership	0.007	0.156	0.045	699	0.964
Random-effects					
var (U_0)		0.015			
ρ (U_0)		0.080			

Note Dependent variable: individual resilience. The total number of surveyed individuals was 733. The total surveyed communities were 11. var (U_0) is the variance component at the community level. ρ(U_0) is the proportion of variance at the community level to the total variance

$* = p < 0.1$, $** = p < 0.05$, $*** = p < 0.01$

[a]If we change the assumption on calculating the proportion of the gated community out of the total area of community, the sign of the coefficient is changed. So in this research, "gated of the community" is more like a statistically significant factor to individual resilience. Because the whole research focuses more on factors on individual level, so in the following part, factors on individual level is deserved more explanation and exploration

level. Based on the analyses, the influence of personal perceptions to resilience is explored after considering the influences of different communities with HLM.

6.3.1.1 Associations Between PE and Individual Resilience

Personal efficacy is a dominant factor in individuals' behavior toward resilience. In this research, three factors reflecting personal efficacy are identified, as shown in Table 6.8. Based on the regression results, "leadership" and "public participation demands on related urban planning and management" have significant influences on changes in individual resilience performance. When the measurements of these two variables (shown in Table 6.7) are combined, the model suggests that the ones who are confident in their leadership and have high demands on related urban planning and management are more likely to have high individual resilience performance level.

6.3.1.2 Associations Between PRP and Individual Resilience

Personal risk perception might stimulate residents to behave in a resilience-oriented manner in ways as described in previous researches. In this part of the analysis, the explanatory factor comprises four variables: "feelings about urban flooding," "concerns about children during urban flooding," "concerns about the elderly during urban flooding," and "sensitive to urban flooding." The results of the regression analysis (Table 6.8) indicate that "concerns about children during urban flooding" is a significant factor in the variance of individual resilience performance. The results show that this factor is positively related to the changes in individual resilience. Along with the coefficients in Table 6.8 and the measurements of variables shown in Table 6.7, this finding suggests that respondents with higher concerns over the safety of children, especially those having children themselves, are more likely to demonstrate higher individual resilience.

6.3.1.3 Associations Between SSP and Individual Resilience

Social support perception may likewise influence individual resilience. In this research, two types of support were examined: "help and support from the government" and "help and support from their workplaces." The regression results suggest that "help and support from their workplaces" and "help and support from the government" are significant factors in the changes of resilience performance for individuals. When the measurements of these two factors are combined, we can conclude that help and support from governments and their workplaces positively influence individual resilience performance. It suggests that the residents who received help and support after urban flooding disasters can cope with and adapt to urban flooding better.

6.3.1.4 Associations Between PEOF and Individual Resilience

In this research, four questionnaire items identify Gongming residents' previous experiences of urban flooding: "number of days worked per week," "the primary means of commuting," "experienced urban flooding," and "losses during urban flooding." The results of the regression model (Table 6.8) indicate that "experienced urban flooding" and "losses during urban flooding" have significant influences on the performance of resilience. Both factors positively influence the variance of the dependent variable. The model suggests that those who had suffered from urban flooding and lost things during urban flooding, such as property or living and production materials, are more willing to behave towards resilience than those who have not. This finding indirectly demonstrates residents' weak awareness of the dangers of urban flooding unless they have been adversely affected.

6.3.1.5 Associations Between PEP and Individual Resilience

Personal perceptions of the quality of physical environment of living zoens are measured by respondents' "amount of construction projects in progress around communities," "areas of ecological lands around communities," "perception of the living environment management in the community," and "perception of waste management created by natural hazards of communities." These items closely relate to the perceptions of individuals and further influence their actions to behave in a resilience-oriented manner and could significantly contribute to resilience through individual actions. Therefore, these variables are in the regression models as independent variables to assess their linear relationships to the performance of resilience. The results of the regression models (Table 6.8) find that "living environment of communities" and "waste management created by natural hazards of communities" are significant. Combining the coefficients in Table 6.8 and measurements of Table 6.7, the results suggest that residents who have better perception of the quality of physical environment, such as clean environment and efficient waste management, are more likely to have high level of resilience performance.

6.3.1.6 Associations Between SCP and Individual Resilience

Social cohesion perception included variables measuring "mutual trust during times of difficulty," "mutual help during times of difficulty," "perception of life in your residential area," "*hukou* status," and "housing ownership."

The results (Table 6.8) indicate that "mutual trust during times of difficulty" is significant to changes of the dependent variable. Moreover, the variable "perception of life in your residential area" is also a factor explaining the changes in individual resilience. The results suggest that these two factors positively influenced the variance of self-organization. Overall, residents with a sense of well-being in their communities, friendly neighbors, and the willingness to trust them are more likely to have high level of resilience performance.

6.3.1.7 Individual Perceptions in Common and Individual Resilience

Based on the previous regression results, we conclude that individuals with high resilience usually had more positive perceptions of the quality of physical environment, social cohesion of their communities, self-efficacy, personal sensitivity to the disaster risks and the experience of urban flooding as well as perceived help and support from governments and industries. These perceptions can be subdivided into three groups: (1) personal efficacy, i.e. a personal judgment on how well one can execute actions required to deal with prospective situations, including PE; (2) disaster awareness, i.e. personal perceptions from their own experiences about urban flooding, including the PRP, SSP and PEOF; and (3) environment cognition, i.e. personal perceptions of the quality of living environment, including the social environment (SSP) and physical environment (PEP).

Personal efficacy includes the factors that residents believe they could offer leadership in times of difficulties and that they have high demands to participate in management and planning for strategies to cope with and adapt to urban flooding, can facilitate the residents to behave in a resilient-oriented manner. Therefore those who have high personal efficacy are more likely to have high individual resilience, and higher ability to cope with and adapt to urban flooding.

High disaster awareness usually makes individuals more resilient. In the study area, the main factors nurturing residents' disaster awareness are that concern over the safety of children during urban flooding, have suffered serious urban flooding, have lost personal belongings or even their residences, and have gained help from their workplaces or governments. Based on the results, we can summarize that only the ones suffered serious urban flooding and lost something during urban flooding are more willing to cope with and adapt to urban flooding. To conclude, in this study area, people with high individual resilience currently are always passively driven to improve their disaster awareness instead of actively improving disaster awareness.

Good living environment cognition also enables individuals to show high resilience. In the study area, communities with a clean physical environment, e.g. efficient garbage collection and a better social environment with higher mutual trust and well-being perceived by residents, tend to have higher level of individual resilience performance.

6.3.2 *Community Attributes and Individual Resilience*

Models shown in Table 6.8 suggest that "gatedness of the community" affects individual resilience performance. Results showed that the survey respondents who live in gated communities are more likely to have high individual resilience.

Gated residential communities emphasize the boundaries between the privately-managed and publicly-managed areas (Le Goix, 2006). Gates play a role in fragmenting and segregating social groups. As a product of urban development, they enclose spaces and select residents through marketing and price (Aulia & Ismail, 2016; Eves, 2007; Le Goix, 2006). Most gated communities are managed by private corporations, and the management is relatively independent of municipal incorporation. In China, gated communities symbolize good community life, and satisfactory urban services (Pow & Kong, 2007; Shamsudin et al., 2016). Therefore, these properties are usually more expensive to live in. In this sense, social inequality exists in different types of communities in that residents have different socioeconomic status and enjoy different levels community management and services. As the above-mentioned findings, residents of gated communities and non-gated communities show different individual resilience performance in Gongming. Based on the self-reported survey data, only around 50% of all residential zones surveyed are gated. In terms of gatedness, all commercial properties are gated communities, while small property housing can either be gated or non-gated. According to the Detailed Planning of Gongming, around 80% of the surveyed people live in small property housing instead of commercial property. It means the dwellings of the other 50% people living in open communities are residing in the small property housing. Based on the regression results, we can conclude that the relatively poor ones living in the small property housing in open communities are more likely to have low individual resilience. This social group should be the focus in the attempts for individual resilience performance improvement.

In summary, the results of HLM prove that in Gongming, both the social environment as well as the personal perceptions of the quality of social environment can influence individual resilience. For the physical environment, the physical environment itself does not play a dominant role in influencing behavior towards desired resilience, but it is the differences of the people's perceptions of the quality of physical environment of different communities that have caused the variances in resilience. It is the empirical foundation for the next chapter to identify place-responsive measures to change human behavior towards the desired level of resilience performance.

6.4 Discussion

Exploring the characteristics of individual resilience for residents of different communities and measuring the factors associated with individual resilience is challenging. This chapter explores the influencing and limiting factors of individual resilience with a bottom-up method to explore the individual attitudes and community attributes on individual resilience performance. The results support the answers for the second research question of this research: What were the determinants (People-Disaster-Place-Society) of individual resilience? This empirical case explored the factors that reflect the ways in which personal perceptions of urban flooding, the physical environment and the management of their communities (People-Disaster-Place-Society) contributed to individual resilience performance. It has tested the second hypothesis of this research that individual perceptions of urban flooding, living place and living society can influence individual resilience. These findings can help and provide support for the urban management to focus on nurturing resilience because they help to identify the characteristics of the people who would benefit the most.

6.4.1 Active Coping with and Adaptation to Urban Flooding (People)

Self-efficacy positively influences individual resilience performance. In our study area, among the three perceived factors, people who have the confidence to lead others out of difficulties and participate in the urban flooding management of their communities show high individual resilience. It can be inferred that their resilience-oriented behavior is based on their own willingness to do so. Their behavior is facilitated over time and by their attitudes to life and work. In summary, the ones who have high self-efficacy show higher resilience.

6.4.2 Passive Coping with and Adaptation to Urban Flooding (People-Disaster)

Personal willingness to cope with and adapt to urban flooding can be passively driven by personal risk perception, previous experiences of urban flooding and social support perception. Low disaster awareness, including an insensitive personal risk perception and less experience of disasters or drillings, limited individual resilience. Interactions between people and disasters can be passively driven by their personal perceptions of natural disasters. The surveyed residents with higher concerns over children's safety in Gongming were more likely to have strategies for coping with and adapting to urban flooding, and those who had lost personal possessions in previous urban flooding events were more likely to show high individual resilience. Furthermore, people who had received assistance and support in natural disasters also show high individual resilience. In sum, personal perceptions of natural disturbances can passively drive personal needs to show resilience-oriented behavior.

6.4.3 Physical Environment and Individual Resilience (People–Place)

Place means the physical environment. Although the regression results suggest the variations in the physical environment of different communities cannot distinguish the residents with high individual resilience from the ones with lower resilience, personal perceptions of quality of their living places can influence individual resilience performance. To conclude, the interaction between people and place affected resilience performance. In Gongming, the people who show high individual resilience are most satisfied with the quality of physical environment. The ones who have positive perceptions of their living environment, such as a clean and tidy physical environment, are more likely to show high individual resilience. Furthermore, in terms of waste management, residential areas with unsanitary or poor waste management can negatively influence the personal perceptions and thus their behavior in a resilience-oriented manner.

6.4.4 Social Environment and Individual Resilience (People-Society)

Better social environment, e.g. gated community with professional housing managers, can positively influence individual resilience. At the individual level, social environment is embedded in residents' perception of social cohesion of their living communities. For residents of Gongming, a sense of well-being in their residential areas could influence individual resilience. In addition, residents who believe in and trust their neighbors and the managers of their communities always show resilience-oriented behavior. It can be concluded that the personal perceptions of the social environment of their communities explained changes in the performance of resilience. Based on these results, we conclude that both the social environment and personal perceptions of the quality of social environment of their communities are a very important explanatory factor of the differences in individual resilience.

6.4.5 Low Individual Resilience in Gongming: People-Disaster-Place-Society

Chapter 5 found that the average score of residents in Gongming only belongs to a moderate level of individual resilience. It means at least half of the residents of Gongming have insufficient resilience to urban flooding. Based on the results of HLM, we conclude that compared to other residents, only those with high individual resilience had positive or sensitive perceptions of urban flooding, the physical environment and the social environment of their residential areas.

Based on the data from the survey, although 46% of the surveyed residents have suffered urban flooding, only 26% have lost personal belongings and around 20% of the selected residents have received assistance. Around 42% of the selected residents do not think the urban flooding is dangerous. In summary, many people in the surveyed area have low disaster awareness and low individual resilience. Combining the data gathered during the fieldwork in Gongming with the spatial data from the Gongming statutory plan, we found that about 50% of the residential areas were open urban villages with inefficient or even no housing management at all. Some of the houses were illegally built with substandard infrastructures. The figures (Appendix D) from the fieldwork show that living environment and waste management were usually quite poor in the residential areas, particularly for residents in the Citianpu, Genzhuyuan, Yulv, and Tianliao communities. Under this situation, it is natural that most people in Gongming have a poor perception of the quality of physical environment and social environment. We conclude that, even among those with high individual resilience, poor social cohesion in the living environment and immature, or even nonexistent, housing management prevented them from realizing the desired individual resilience to urban flooding.

To conclude, at the individual level, low personal efficacy, low disaster awareness, and poor quality of environment cognition among individuals limit the resilience performance of Gongming residents. Six groups of variables reflecting the personal perceptions of urban flooding, physical environment and social environment (People-Disaster-Place-Society), totaling 11 groups of associations that explain individual resilience performance, are depicted in Fig. 6.1. In the figure, red represents PE, orange represents PRP, blue represents SSP, green represents PEOF, purple represents PEP, and yellow represents SCP. The lines indicate the connections between the items with the direction of influence, and the widths indicate the strengths of the effects when each influencing factor is equally treated regarding their P-values, which is applied to assess their explanatory power.

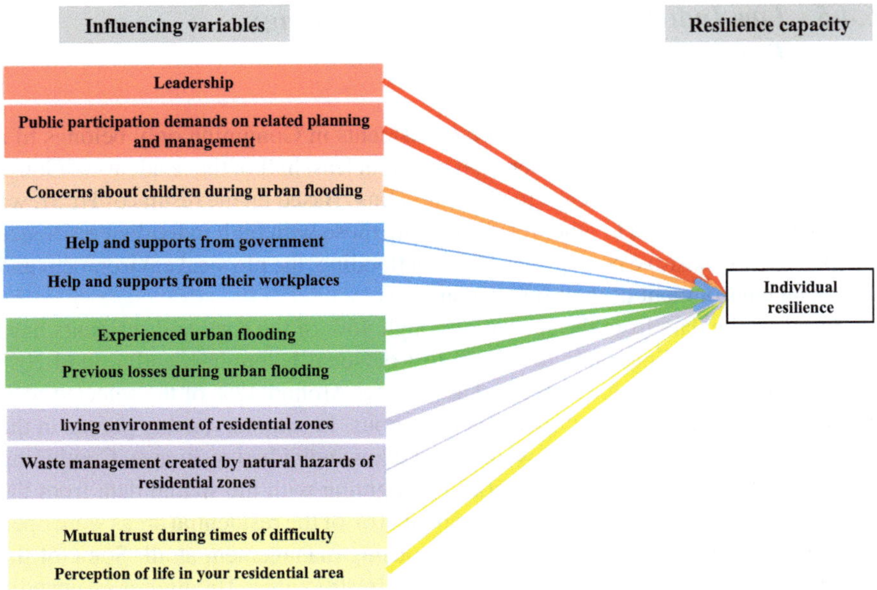

Fig. 6.1 Associations between influencing factors and individual resilience. *Source* Author

Half of the residents in Gongming lack individual resilience without caring about improving their social resilience to urban flooding. This may be attributable to the fact that these residents are already struggling to satisfy their basic living needs, not to mention developing personal efficacy, disaster awareness and caring about living environment. However, due to the increasingly uncertain and extremely heavy rains associated with climate change, these people will be in danger when there is a disastrous rainfall and urban flooding. As reviewed in Chap. 2, individual resilience performance can be pushed towards the desired direction to cope with and adapt to urban flooding. The political arrangements for disaster management as an external intervention can nurture resilience. If the resilience-oriented strategy, Sponge City Initiative can nurture individual resilience should be identified based on the findings of this chapter.

6.5 Conclusion

This chapter improves our understanding of the factors that influence individual resilience performance. This study is conducted at the individual level but also considers the influences from the community level. The analytical results find that personal perceptions of urban flooding, quality of physical environment, and social environment do influence the performance of individual resilience. On the community level, the regression analyses also suggest that better social environment can positively influence individual resilience performance. The knowledge generated by this study will allow measures to be developed for vulnerable communities with the least resilience.

Chapter 7
Nurturing Individual Resilience Through Urban Management

7.1 Introduction

In Chap. 5, we discussed that the individual resilience performance is insufficient to reduce the impacts of high vulnerabilities to urban flooding in Gongming. In Chap. 6, we identified 11 variables covering six groups of factors reflecting personal perceptions that limit individual resilience performance in Gongming. We also found that social environment on community level in the form of gated community increases individual resilience. The identification of these factors does not only confirm that the personal perceptions of urban flooding, physical environment and social environment can influence individual resilience performance, forming the People-Disaster-Place-Society-Resilience relationship networks, but also provide the empirical foundations for place-responsive urban management to improve individual resilience performance. The conceptual review in Chap. 2 points out that nurturing individual resilience is closely related to the learning paths of individuals, social groups, and governments. Low individual resilience suggests problems related to learning paths in Gongming; however, we still do not understand the degree to which the urban management can nurture human behavior towards resilience. Therefore, whether management can compensate the deficiency of the learning paths needs to be examined.

This chapter addresses the third group of research questions: (3) Whether the strategic measure of Sponge City Initiative could help nurture individual resilience? And what were the place-responsive measures for individual resilience nurturing? And to test the third hypothesis that individual resilience cannot be nurtured through resilience-oriented measure, i.e. Sponge City Initiate from central government of China. To conduct this research, a resilience-oriented strategic measure—Sponge City Initiative from central government of China is examined. As a new initiative aiming to reduce the problems of urban flooding, Sponge City Initiative can reflect

J. Song, *Individual Resilience to Urban Flooding and the Implications for Urban Management*, https://doi.org/10.1007/978-981-97-0546-7_7

the latest and most advanced urban flooding management.[1] This new initiative also provides the possibility for us to examine the mechanism for urban management from inception to completion. This chapter consists of four analytical sections. In the first section we examine if Sponge City Initiative can activate self-learning, considering its potential to improve disaster awareness regarding personal perceptions of urban flooding. In the second, we examine if Sponge City Initiative can activate collective/mutual learning, considering its potential to improve social environment of Gongming regarding personal perceptions of social environment. We then examine if the Sponge City Initiative, which is formulated based on the learning of governments, have the potential to improve the ecological infrastructures regarding personal perceptions of physical environment. Finally, based on the previous analyses, place-responsive measures that can improve individual resilience are identified.

7.2 People-Disaster and Sponge City Initiative

In Chap. 6, we concluded that personal perceptions of urban flooding affect individual resilience performance. Most residents of the research area show low disaster awareness of urban flooding, and this leads to low individual resilience. Here we identify impediments to residents' self-learning to improve their disaster awareness. Based on social learning theory, we believe that political forces can influence what and how people learn, as well as human behavior towards coping and adaptation to natural disturbances (i.e. resilience nurturing). Therefore we need to explore whether the political initiative of the sponge city can make up for the deficiency or failure of self-learning. This section examines and explains the reasons for the failure of self-learning and the extent to which the Sponge City Initiative can make up for this failure.

7.2.1 People-Disaster on Self-learning

Chapter 6 suggested that factors closely related to disaster awareness comprise (1) PRP, embedded into the concerns about children during urban flooding; (2) PEOF embedded into factors of previous experiences on urban flooding and their losses during urban flooding; and (3) SSP embedded into factors of help and support from governments and their workplaces. These are selected for correlation analyses on the demographic characteristics of selected residents, which can help inform us the attributes of the group of people. The models shown in Table 7.1 suggest that the low-income, single men appear to have low disaster awareness in Gongming. Interviews were conducted to verify these characteristics.

[1] http://opinion.people.com.cn/n1/2016/0921/c1003-28728354.html.

Table 7.1 Results of correlation analyses between self-learning limiters and resilience-related personal characteristics

Model predictors	Model 1 ρ (N = 733)	Model 2 ρ (N = 733)	Model 3 ρ (N = 733)	Model 4 ρ (N = 733)	Model 5 ρ (N = 733)
	Concerns about children during urban flooding	Help and support for the government	Help and support for the workplaces	Experienced urban flooding	Previous losses during urban flooding
Gender	– 0.100**	0.010	0.072	– 0.068	0.173*
Marital status	0.308**	0.010	0.042	– 0.087*	0.063
Age	– 0.134**	– 0.001	– 0.068	– 0.002	– 0.025
Occupation	– 0.051	– 0.011	– 0.010	– 0.097**	0.044
Shenzhen *hukou*	– 0.026	0.045	0.002	0.016	0.043
Education level	0.011	– 0.031	– 0.030	– 0.034	– 0.025
Housing ownership	0.031	0.067	– 0.021	0.041	– 0.019
Housing types	– 0.088*	0.012	0.015	– 0.082*	– 0.004
Income	0.020	– 0.026	– 0.031	0.165	– 0.107**

Note ρ is the correlation coefficient between the two variables
* $= p < 0.05$, ** $= p < 0.01$, *** $= p < 0.001$

A 25-year-old married female said, "I think it is a little dangerous, especially since I have a little girl, only three years old. I saw the news that there was a child who fell into the open well when there was heavy rain." (Residents 7).

A 25-year-old male commented, "I haven't suffered from the effects of urban flooding." (Residents 4).

I asked, "Have you ever seen or suffered waterlogging on the road after a heavy rain? The waterlogging is called urban flooding. Urban flooding doesn't mean flooding."

He answered, "Oh, yes, I usually suffer from waterlogging on the roads in the rainy season of Shenzhen. But I do not think it is dangerous to us." (Residents 4).

Another young man remarked, "I do not care about the heavy rain and urban flooding. Landslide in Guangming Sub-district of Shenzhen[2] cannot alert me so I do not think I need to do any preparation for heavy rain and urban flooding." (Residents 6).

The statistical data and interviews suggest that there is a population in Gongming who are mainly single and young, living in urban villages without families to worry about during heavy rain and urban flooding. They are mainly the males who

[2] Secondary disaster associated with heavy rain, which occurred at Shenzhen, China, on 20 December 2015 (https://en.wikipedia.org/wiki/2015_Shenzhen_landslide).

show negative attitudes towards disasters, having low disaster awareness and passive attitudes about adapting to urban flooding.

Literature states that self-learning is blocked by people's reluctance to alter their behavior and perceptions in anticipation of, or following, an extreme event (Bohensky & Leitch, 2014). There are no personal or external motivations for the residents of Gongming to improve their disaster awareness toward urban flooding, even though these flooding can be dangerous. The residents of Gongming are usually peasant workers with low income and limited education. They may have to work for a living, which limits their ability and motivation to understand urban flooding, especially if they are single and without children. There are awareness barriers for these people. This population is more likely to have low disaster awareness, forming a new type of imbalance in information flow. Considering that imbalance in information flow has existed, interventions that aim to change individual behavior may have to be made through structural and informational measures. "Informational strategies are aimed to change the motivations, perceptions and norms, whereas structural strategies are aimed to change the circumstances under which behavior choices/decisions are made" (Soetanto et al., 2016). Considering imbalance in information flow, it is suggested that the informational measures of urban flood management should be disseminated to all, but tailored to be more effective for residents who are low-income, or have negative attitudes towards disasters.

7.2.2 Sponge City Initiative for Resilience Nurturing in Terms of People-Disaster

This section will examine if government interventions will give external motivation to reduce awareness barriers for this group of people showing imbalance in information flow on improving awareness of disasters.

7.2.2.1 Demonstration Projects for Reducing Imbalance in Information Flow

There are 18 completed sponge city construction projects in the Guangming New District, with their spatial distribution shown in Fig. 7.1. Low Impact Development (LID) is the embryonic form of sponge city projects, and these 18 cases are essentially LID projects. Since these cases have been completed, the special agency of Sponge City Initiative of Shenzhen encourages residents to visit the pilot projects to improve their disaster and water-saving awareness.[3] This is a good way to disseminate information on disaster awareness and explain why the city spends so much money on sponge city construction.

[3] http://www.szwrb.gov.cn/ztzl_78021/bmzdgz/hmcsjs/gzrw/201612/t20161203_5599798.htm.

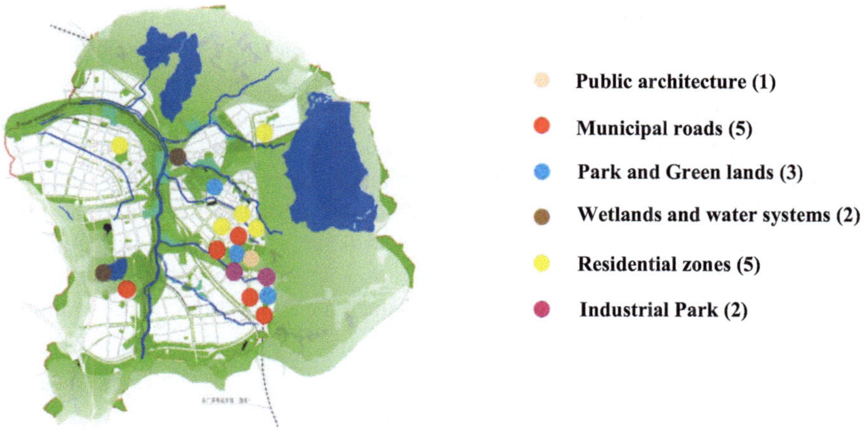

Fig. 7.1 Spatial distribution of LID projects in Guangming New District. *Source* Special urban planning for sponge city construction (2016)

To determine the educational effects of the visits to the pilot projects, we used the search engine website of Baidu to look for related news. "Guangming New District," "Sponge city" and "Visiting" were the keywords searched. The results are shown in Table 7.2.

Based on the search results, we found 52 pieces of news about the visits to the pilot LID projects. The visitors tended to be government officials, developers, planners, constructers, scholars, and NGOs. The sponsors tended to be governments,

Table 7.2 News information summary based on web data

News No	Visiting aims	Sponsor	Main visitors
3	Disaster awareness improvement and environment protection	Schools of Shenzhen	Students, teachers, and planners
2	Sponge city construction learning	Other city governments	Governments
12	Technology for sponge city construction	Shenzhen governments	Planners, architectures, engineers, and governments
6	Policy making	Shenzhen governments	Governments
9	Methods and management exploration	Scholars or NGOs	Scholars, NGOs, and governments
17	Sponge city propaganda and construction aim	Media	None
3	Financing for sponge city construction	Shenzhen governments	Governments, industries, such as Tencent and NGOs

Source News from search engine websites of Baidu

scholars, NGOs, and the media. The visits were meant to publicize the Sponge City Initiative, making policies and identifying construction strategies, including technologies, financing, and management. Only three news articles mentioned students visiting the pilot projects. We found that most of the visitors were associated with the government, and that the purpose was to design and build a sponge city. This would complete the sponge city construction target set by the central government in terms of hardware without focusing much on giving external motivation for nurturing the disaster awareness of the residents. Most residents, even those in Gongming, are unaware of the sponge city projects in Guangming New District, let alone the reasons for the construction and how they can help reduce the dangers of urban flooding. As Planner A remarked:

> The main aim of the governments of Shenzhen in building a sponge city is to play a leading role for sponge city construction in the other cities, in other words, the officials focus more on the political achievement. (Planner A)

This suggests that the new agency focuses on packaging and building sponge cities as quickly as possible, without sharing the concept with and building disaster prevention awareness among residents. Data from the questionnaire shows that 573 out of 733 residents (78%) do not know what the sponge city is, and of the other 22%, only 10% have only a general idea. This means only 2.2% of the selected residents have heard of the Sponge City Initiative. None of the local residents interviewed knew that there are 18 sponge city projects in the Guangming New District. Here are some statements from the residents interviewed (Table 7.3).

A sponge city project is a public infrastructure provided by the government, so public services, such as disaster awareness improvement, should also be the responsibility of the state (Grimsey & Lewis, 2002). Providing public infrastructure without the accompanying public services fails to be "tailored" to particular groups, which exacerbates the imbalance in information flow in improving disaster awareness (Soetanto et al., 2016). Such government-led activities depend on the attitude of the government itself towards these projects. These attitudes are a trade-off resulting from differences in incentives (e.g. political achievement, market profits, and services to the residents). Governments portray themselves as having considered all incentives; however, they might concentrate only on what they see as the most urgent. During this process, social services are always ignored or given up by the state-centric model (Bushley, 2014; He, 2003), because the effects of social services are not easily identified and evaluated. This means this incentive is not as direct and evident as the other two, which leads to the government's failure in providing external motivation for disaster awareness improvement for local residents.

Table 7.3 Statements on the visiting sponge city cases for the residents in Gongming

Statements	Attitudes towards improving disaster awareness
I have never heard about the sponge city, and I do not want to learn about it either. (Residents 1)	Low demands
Really? Is there a sponge city demonstration project in Guangming New District? I have never heard about it. Can it help to reduce the effects of urban flooding? Why is it built there, and I think that place has lower population density, and it seems not as urgent as my place. (Residents 2)	Willing to but not *very* high demands
I have never heard about it, but I think I can guess the rough meaning behind it. (Residents 9)	Willing to but not very high demands
I think I saw some news about it on the TV, but I cannot remember it exactly. (Residents 12)	Willing to but not *very* high demands
I have never been invited to visit the demonstration projects. In fact, I do not want to visit it either. (Residents 15)	Low demands

Source Summarizing the responses from interviewees

7.2.2.2 Smart Sponge City for Reducing Imbalance in Information Flow

Sponge city construction in Shenzhen involves participation of High-Tech company to help information dissemination. As an indigenous and high-end technology company, Tencent Company, with the Tencent Foundation, integrates with the urban planning department of Tencent, Tencent Cloud, and Wechat City Service, will provide a public platform for technological support and information transmission for the sponge city. This information not only advertises advanced resilience, but also focuses on improving disaster awareness, accuracy, the timing of early disaster warnings, and self-protection. This platform can also provide real-time monitoring and understanding of extreme weather and urban flooding based on integrated data, such as weather, transportation, and personal location data. It is based on the Wechat platform, because most residents in Shenzhen use Wechat City Service. This service could reduce the risks of urban flooding to a certain extent.

Returning to Gongming, the data collected from the questionnaire shows that 31% of the sample residents are not interested in learning about sponge city or about urban flooding. Of these residents, 39% of them have an income lower than 3000 CNY/month, and 85% of them have income lower than 7000 CNY/month (the average wage in Shenzhen is 6753 CNY/month). Of these residents, 76% are not educated above high school. Data from the selected sample also shows that 27% of residents do not want to check weather reports and similar products. Among them, 30% earn less than 3000 CNY/month and 86% less than 7000 CNY/month. 26% have no more than middle-school education, and 69% of them have not attended college or university. Statistical data suggests that the informational measure provided by Tencent is beneficial to most residents; however, some residents we want to tailor for may not use them.

Literature suggests that efficient intervention of urban flooding management should be tailored for different demographics, particularly people with the least individual resilience (Berkes, 2007; Tschakert & Dietrich, 2010). Urban flooding management should not be limited to the majority of residents who have the least effort, but also focus on the groups whose individual resilience are extremely low. If there is a disaster, those with low individual resilience will suffer the most.

In conclusion, driven by the incentive of political achievements, the centralized and hierarchical government tends to ignore the provision of social services, but focuses on public infrastructures which can be identified and evaluated easily. Present services from the government and industries still focus on offering services to most of the population with the least effort. Social services should be tailored for vulnerable groups, particularly for vulnerable areas (Soetanto et al., 2016). In addition, nurturing the resilience of residents should be the responsibility of the government. Under the current management regime, government-led production cannot make up for the deficiency in resilience nurturing, especially for residents facing imbalance in information flow.

7.3 People-Society and Sponge City Initiative

In Chap. 6, we found that a constraint for individual resilience is the perception of poor social environment under poor housing management and immature neighborhood management for safeguarding residents. This suggests that the residents of Gongming cannot learn from each other to improve their individual resilience. Political forces can influence what and how people learn, and affect them to behave in a resilience-oriented manner. We need to understand if the strategic measure of Sponge City Inititiative can make up for the failure of collective learning in resilience nurturing.

7.3.1 People-Society on Collective Learning

Collective learning is a key component of social learning to nurture resilience-oriented behavior. "Understanding the ecosystem processes and how to manage them seems to be a progression of social-ecological coevolution" (Olsson et al., 2004, p. 77). Collective learning is an opportunity for people to accumulate and learn new knowledge based on others' experiences, and share their own information, measures, and opinions. Changes and successful adaptions are embedded in a deeper level of values through social groups' debate and decision-making processes, forming appropriate measures for dealing with ongoing change (McIntosh, 2000; Olsson et al., 2004; Tschakert & Dietrich, 2010).

In the previous chapter, we identified some limiting factors, including mutual distrust that reflected poor collective learning at the community scale. Communities with mutual distrust or poor well-being perceived by their residents usually do not have high quality of social environment. Social environment are the foundation for residents to learn to behave towards desired resilience. Based on previous analyses, we understand that a poor social environment of Gongming blocks the collective learning path of community residents. This fails to nurture resilience, so we need to understand why the personal perceptions of living in the community are so poor.

As mentioned in Chap. 2, formal and informal stakeholders like residents, community managers, and some social groups, play critical roles in cooperating with the government, facilitating social environment for collective learning from neighbors (Olsson et al., 2004). Literature suggests that in western countries, the main forces for resilience are grassroots organizations (Chandra et al., 2010; Rose, 2004). However, in Gongming, this does not seem to be the case. There are some special factors in our research area. Based on the statutory plan map of Gongming, spatial data shows that limited rights housing (urban villages) accounts for 80.11% of its residential land, and most of them are rental homes occupied by migrant workers. The supporting infrastructure of these communities is inadequate with poor management, inducing many safety hazards. Among these houses, there are still rural lands that have not been transformed into urban residences, and they are privately owned and constructed by illegal developers. Clan and blood ties pose challenges to property management. In managing communities, the Rural Shareholding Cooperative (RSC) appears to have a more important role in motivating collective action (Ito et al., 2016; Liu, 2015). However, these collective actions exclude the migrant population, further limiting an efficient collective learning process. One migrant worker remarked that:

> I have lived in one apartment in a privately owned small property right housing for more than two years. Before that, I lived in a flat with even poorer conditions, which also belongs to privately owned small property right housing. I have been in Shenzhen for eight years, and the two communities I have lived in are single buildings without housing management. I have heard that the owner of the building I am living in and the owners of the buildings around are brothers or cousins. I have not participated in, or been invited to participate in, any communal or community activities. I think the locals may have this type of organizations. (Residents 11)

In conclusion, complex interactions of RSC with different formal rules (e.g., administrative restructuring, rural planning) and informal constraints (e.g., clans, rentier) make social environment exclusive, limiting the potential for collective learning particularly among the non-native population.

Community management also influences the social environment of communities (Leyden, 2003). However, in Shenzhen, the rights and obligations of sub-district committees and the communal committee are usually ambiguous, leading to the shrinking of responsibility for the provision of social services to residents. There is lot of work that community management should take responsibility for. The community is characterized as being "many organizations with many heavy tasks," and assumes at least 60 governmental tasks. More than 50 of these are specific rectification works, but communal committees are understaffed and underfunded. Most of the activities remain at the stages of precepts, meetings, applications, summaries, and reports. A community worker remarked:

> Information gathered, data statistics, and indicators for performance appraisal by the communities are a lot to hand over to the governments. Every month, we need to submit at least 13 categories, 56 classes, and 358 indicators to the governments. (Residents' committee B)

The heavy workload and pressure, with an overworked labor force, led to a mode of administration that neglected social services, social insurance, and cultural and spiritual development. It is the cause of the failure of collective learning. Besides, it becomes harder for grassroots government to provide social services, because the informal relationships, such as clan and rentier are strong, which restrict collective learning at the community level. We can therefore conclude that collective learning is a failure in Gongming. We need to identify if government-led learning, such as the political behavior associated with sponge city construction, can make up for the failure of collective learning.

7.3.2 Sponge City Initiative on Resilience Nurturing in Terms of Poor Social Environment

Based on previous analysis, the poor social environment of community residents has led to the failure of collective learning and thus low individual resilience. Therefore, this section will examine if interventions from governments embedded into sponge city construction will improve the social environment in community management.

Besides the special planning or news on websites related to sponge city projects, we have not mentioned that there are specific institutional measures for working on the improvement of social environment of residents. The management staff remarked that, until now, they have not received any government information about the Sponge City Initiative. The present measures are the same as the previous for coping with urban flooding, such as sewage system maintenance, potential safety hazard checking and preventive measures before heavy rain. It is hoped that social organizations, such as NGOs, will work on it. Literatures state that in western countries, NGOs play an important role in reducing imbalance in information flow and contributing to social environment (Clark et al., 1998; Lyon & Maxwell, 2008). They are encouraged to work with the communities that they serve to educate and support them at all stages of disaster management.

7.3.2.1 NGO for Improving the Social Environment

For the resilience-oriented measure of Shenzhen, the Sponge City Initiative attracts and invites NGOs to participate. Two well-known NGOs, the Paradise International Foundation and the Nature Conservancy, cooperate with the government on the sponge city projects.

Paradise focuses on exploring a sustainable financial model for sponge city construction in Shenzhen, while the Nature Conservancy helps the government formulate policies and management framework. This company has nearly 10 years of experience in developing policy and management measures for similar projects in sponge city construction. Developing policy and management measures are a principal component of sponge city projects. Policy is the plan for the project, while mature management guarantees its smooth operation. Although there is a financial grant from the state, sustainable sponge city construction requires healthy and stable financing, not only in the following years but also for all the future developments in Shenzhen. Possible and sustainable financing models are therefore pressingly needed and important for Shenzhen's sponge city construction. Participation of NGOs helps to solve these two critical problems in sponge city projects.

Compared to the roles of NGOs in western countries in similar projects with similar aims, NGOs in Shenzhen, China, seem to focus less on social services for individuals. There are several reasons for this. Firstly, China is a centralized society, governed by a one-party autocracy. Under this regime, the government has control of everything, especially social organizations, which make it difficult for NGOs to take any action. NGOs have no independence, and they do not act as a third party to improve social environment. For this reason, NGOs in China are operated under a typical top-down model, which means that they serve the government. A staff member of an NGO remarked:

> Last month, we held a forum on water resilience. We invited lots of government officials, scholars, representatives from the industry, as well as students in colleges and universities. But we did not invite the residents, for we think the public engagement should be the responsibility of the government" (NGO 1)

Financial constraints limit them in performing what they want to do. In China, the main financing organizations for NGOs are the government and foundations. NGOs, therefore, always cooperate with government and foundations to perform social services instead of serving residents directly. Most of the residents of China cannot or do not want to invest in this type of service. A representative of an NGO said that, "in at least 10 years, we cannot hope that the residents of China will raise the money through crowdfunding in exchange for services directly serving them." (NGO 2).

Some international funds want to invest in Chinese grassroots NGOs. However, once these groups receive international funding, the government will become suspicious, and often restrict their behavior, fearing conspiracies from foreign countries.

Some GONGOs with official backgrounds or support often threaten the grassroots NGOs since they have substantial funding from the government, allowing them to achieve high self-development capacity and attract talents. GONGOs always serve the government directly, which means that they are not independent organizations. This can explain some of the differences between foreign and Chinese NGOs.

Though survival is difficult for grassroots organizations, they still have opinions on the services for residents for improving their ability to cope with and adapt to urban flooding. Another member of staff of an NGO noted:

> I think we can do something for the residents, and we can tell them how to deal with garbage effectively, which can help reduce the potential risk of blocking sewage systems. However, it should meet the government's or foundation's present work plan. Now, the government seems to be focusing more on the physical side of sponge city development, and I think it is a little difficult for them to carry out these things in the short term. In fact, we seldom play a dominant role in urban flooding management. But we may try to make suggestions to the government for consideration. (NGO 3)

In summary, a state-centric model (Brenner, 1999; Clarke, 2017; Marks et al., 1996) restricts the vigor of grassroots services, which further decreases grassroots' social environment building. NGOs in China still show financing-oriented characteristic, which means that they serve those who fund them without autonomy. Even though most of them provide social services that are beneficial to residents, their work is still confined to executing the government's plans. This explains why only the NGOs that work for and receive funding from the government can be engaged in the Sponge City Initiative, instead of NGOs that carry out public education for the residents of the local community.

To conclude, the present resilience-oriented measure of sponge city projects has not focused on social environment building. This is a defect of the resilience-oriented measure, which could lead to its failure in combating urban flooding management. Building a sponge city has multiple aims, and urban flooding prevention is just one of them. It is unfair to regard the whole project as a failure, but it has addressed the root causes of urban flooding, due to its failure in building social environment. It is difficult for both the formal and informal grassroots organizations to help advance the present social environment of communities like Gongming. In the formal organizations, there has been excessive administrative work for the staff to deal with the government, which restricts their time and energy for improving social services for residents, particularly migrant workers living in private, illegal buildings. In the informal organizations, the centralized government structure does not allow them sufficient room to perform their role such as providing social services to those with low level of resilience performance, which makes it harder for grassroots to benefit from collective learning.

7.4 People-Environment and Sponge City Initiative

Personal perceptions of natural disturbances and management of Gongming lead to poor individual resilience performance, as reflected in the failure of self-learning and collective learning for nurturing resilience. The strategic measure of Sponge City Initiative as a high-order initiative from the central government cannot compensate the deficiency of self and collective learning of residents in Gongming.

Based on the limiting factors identified in Chap. 6, the personal perceptions to the physical environment of their communities also influence residents' individual resilience performance. As repeated several times before, there are many small property housing in Gongming which are developed with incomplete infrastructure, especially in terms of sewage systems and green spaces. Therefore, the poor perception of residents to the physical environment of their communities limits the Gongming's individual resilience performance.

Since the government is in charge of providing public infrastructure and community planning, what is wrong with the previous government learning that led to the poor physical environment of communities in Gongming? We need to examine if the institutional measure of Sponge City Initiative can make up for the failure of previous government learning on nurturing resilience. Therefore, this section examines and explains reasons for the failure of previous government learning and the degree to which the Sponge City Initiative can make up for the failure of previous government learning.

7.4.1 People-Environment on Government Learning

Social learning is a complex process to aid the adaptation to climate change, build social trust and involve diverse stakeholders. It allows individuals and social groups to negotiate, challenge, and propose new programs, norms and policies. Social learning sectors include individuals, social groups, organizations or institutions and governments. As discussed above, government learning could fail. Since the municipal government is the top-learning sector within a city, there should be a learning structure leading to the present learning problems of the government. Loop-learning theory discussed in chapter two is used here to help judge that loop of learning failed for the government.

Poor sewage systems are a reflection of a poor physical environment, which can lead to an unsanitary environment and reduce the effectiveness of waste management. There are some characteristics of Gongming that lead to the failure of the sewage system during rainfall. Shenzhen is a region of high precipitation and the spatial distribution of rainfall is uneven, with western and eastern parts receiving extremely high amounts of precipitation. Gongming is located in the rainy area of Shenzhen. Intensive extreme weather conditions that last for a long period of time frequently occur in Gongming, challenging its drainage capacity. Although Gongming is rapidly urbanizing, its infrastructure development cannot catch up with the pace of urbanization. At the same time, ecological lands are the amount of ecological land is dropping dramatically, leading to the breakage of natural drainage systems and the lack of functional water bodies. There are many impermeable surfaces in addition to construction projects that have destroyed or blocked the original sewage systems, making an already vulnerable sewage system unable to withstand a single blow. By suffering more frequent, serious urban flooding, governments, including the municipal government, as well as the district governments, are aware of the situation and consider it to be important. Governments have convened several meetings to refine various mechanisms, including emergency evacuation plans, preparation of relief materials, hidden danger identification, and professional rescues. Funding has been collected and prepared for aiding urban-flooding-prone areas.[4]

The government has completed the first learning loop in identifying the problems and providing some political arrangements for it (e.g. the organizational structure, planning, financing, and public education). They also challenge current strategies, as they have found that the present situations are still unchanged. They have realized that ecological restoration and improvement are important. Therefore, they do not only emphasize strengthening the sewage system, but also pursue LID practices for the development in Guangming New District. This has been implemented in about 18 cases since 2003. These works, done through government learning, are shown in Table 7.4.

Government remains at the second learning loop without entering the third loop. According to Table 7.5, the demands of stakeholders for physical environment improvement have not been addressed even when there are LID practices. We can find that the strategies shown in Table 7.4 have not been backed up by political and legal support for their operation. It is necessary for the directives from the central government within a political context to be binding to lower-level governments, enabling efficient management and ensuring effective implementation. Therefore, the missing third loop in the learning process is a likely reason for the poor physical environment built in Gongming. We need to identify if the Sponge City Initiative can make up for this deficiency to lead to improvements in the resilience performance of Gongming ultimately.

7.4.2 *Sponge City Initiative on Resilience Nurturing in Terms of Poor Physical Environment*

Ecological infrastructure construction is a main component of sponge city development, which can improve the physical environment in or around communities to a certain extent. The environment is closely related to the perception of residents and

[4] Process report of urban flooding management in Guangming New District. Source: Guangming Wind, Flood, and Drought Prevention Office (2016).

Table 7.4 Present urban management on urban flooding from governments

Items	Aim	Statements
Organization cooperation	Build city disaster prevention, resistance, and relief systems	Shenzhen Water Authorities has built a hierarchical network named wind, flood, and drought prevention with headquarters for emergency direction and information spread In Guangming New District, grid management is applied, and there are accountable regulation arrangements for people in charge of each flooding-prone area or point, which will be assessed as their performance of officials of governments There are also commissioners to clean, check, and repair the sewage systems at a city scale in a certain circle for avoiding urban flooding Planning (drainage planning, urban flooding prevention planning and sponge city construction planning)
	Spreading information on weather reports Urban flooding simulation and early-warnings Urban flooding real-time dynamic query	The meteorological department is cooperating with the urban planning department and wind, flood, and drought prevention headquarters in Shenzhen, and is improving the accuracy of weather and urban flooding monitoring, prediction, simulation and early warning systems Shenzhen wind, flood, and drought prevention headquarters' offices has set 332 warning signs and water gauges, which helps in alerting and directing pedestrians and vehicles for avoiding losses from urban flooding Shenzhen wind, flood, and drought prevention headquarters' offices cooperates with Meteorological department and has developed apps for mobile phones to receive real-time monitoring results for urban floods
	Communities and communities take responsibility	Under the direction of municipal and district governments, the committee of communities and neighborhoods take responsibility for urban flooding management, including screening residents for potential risks associated with heavy rain and urban floods, e.g. cable leakage in the rain water and so on The committee of communities also takes responsibility for environment protection and garbage disposal

(continued)

Table 7.4 (continued)

Items	Aim	Statements
Financing circulation	Investment	The Shenzhen government introduces Shenzhen work plan on water management and quality management (2015–2020) and invests 81.6 billion on sponge city construction, drainage projects, emergency projects, wastewater treatment, ecological restoration projects, etc. LID (Low impact development) as a construction concept is applied to development and construction projects. More specifically, the construction of green and blue systems, green buildings and architecture, rain water collection systems, and greening of residential areas need to reach to the standard required by governments Industries, such as Tencent, will contribute to building construction with a sponge city. Tencent also launches the public welfare activities for financing LID projects
Public participation	Education and publicity	We can identify some strategies for coping with urban flooding and some self-protection methods provided on websites In some communities, there are also some propaganda to remind residents to help themselves when facing heavy rain and urban floods

Source Author summarized the planning, news and reports on the work of the governments of Shenzhen

can be improved through sponge city construction, in turn improving the individual resilience of residents. This suggests that sponge city construction presumably could improve individual resilience performance. The following section contains a detailed analysis.

7.4.2.1 Sponge Body and Physical Environment

The sponge body, e.g. LID practice, works as a giant water retention unit, and a filtering and recycling structure involving sidewalk, rooftop, street, yard and park in which each droplet leaves as clean and as drinkable as the one that enters (Yu et al., 2015). The sponge body can be a green infrastructure, and includes sustainable, load-bearing, permeable, and breathable roads and pavements, so that rainwater will be stored underneath during rainfall (Qiu, 2015a, 2015b; Yu et al., 2015). Therefore, the sponge body does not only help urban areas absorb rainwater, but also improves the physical environment. We now need to determine if these sponge bodies have been or will be built in Gongming, which could solve the problems of Gongming to a certain extent. The next section will examine this question.

Table 7.5 Demand and deficiency of present urban flooding management of Shenzhen

Items	Demand	Statements
Organization cooperation	Political and legal environment	Until now, there are no laws, rules, or policies to compel a resilience-oriented concept into urban development, such as sponge city construction, and regulate the efficiency and effects of its construction Compared to China, in 2004, New York has created a 4 year-policy named "resilience city construction plan," and in 2007, New York carried out the "Plan for New York." The following year, New York also made plans named "Project evaluation and action plan for climate change." In 2010, this city created a special organization named "urban council for climate change in New York." In 2013, New York enacted "A Stronger, More Resilient New York," a comprehensive plan that contains actionable recommendations for both communities and infrastructure construction
	Efficient and collaborative institutions	"There is no well-complete organization for urban flooding management. There are many departments of governments, as well as industries related to the management, and there is disconnect between rights and obligations among different departments." (Official-D) "Mutual buck-passing existing among different departments once there is disturbance. Different departments are in charge their own work, however, these works have always overlapped, and no one would like to take responsibility for the integration work. It will lead to interrupt or repetition during the disaster management." (Official-A) Governments are still in the leading position and industries are generally invited to cooperate and meet their requirements
	Efficient supervision	"Efficient supervision and monitoring of disaster management are needed. The present supervision of management are as 'the right hand is in charge of the left hand', which means they lack independence between the supervision department and management department of governments in China." (Official-B)
	Timely and efficient disaster feedback	Timely and efficient reports are needed during and following disasters, however, experts and media always enlarge or diminish hazards and convey incorrect or biased information about disaster management

(continued)

Table 7.5 (continued)

Items	Demand	Statements
Financing circulation	Private investment	"We do not want to donate to the management, except if the government requires it." (Residents-9) "We can donate to the management of the sewage system if I am the property owner." (Residents-2) Planning and development as well as redevelopment are profit-oriented and government-dominated processes in China
	Public participation	"I hope I can get some self-saving strategies from community lectures or publicity." (Residents-3) "I haven't been invited to participate in any construction or planning activities by governments, planners or developers." (Residents 4)
Public participation	Public awareness of disaster prevention	475 of 733 residents (around 65%) have not conducted basic common sense to save themselves and others 194 of 733 residents (around 26%) do not have the consciousness to check weather reports and 116 of 733 residents (around 16%) do not think it is important to follow instructions to arrange their commuting based on the suggestions of weather reports 94 of 733 residents (around 13%) are not willing to prepare for heavy rain or do not know how to prepare for probable urban flooding Unexpectedly, 226 of 733 residents (around 31%) do not want to learn strategies to save themselves and others during urban floods. However, after the introduction of some detailed cases related to urban floods, residents who do not think it is serious initially realize it is a serious and dangerous problem
	Living environment improvement	The living environment is still substandard "Littering exists and garbage disposal and waste management are not in time in my living areas, and the festering refuse is piled up like a mountain." (Residents 4) "My living areas have quite a lot of cockroaches, rats and mosquitoes." (Residents-13) Bacteria, viruses, and parasites will run and spread with the floating of rainwater, which will threaten the health of residents. Garbage will also block the sewer under washing by heavy rain, which will increase the risk of urban floods

Source Author summarized based on analytical results of this research, news, literature reviews, and interviews

7.4.2.2 Planned Spatial Distribution of Sponge Bodies in Guangming New District

According to the special planning for the sponge city construction,[5] excluding the 18 completed projects, 51 new projects, which will be completed in the upcoming five years (2016–2010), are either under construction or being planned. Sponge body construction is embedded into the construction of municipal roads, communities, public parks and green lands, civil infrastructure (sewage systems), and disaster prevention and ecological water restoration projects (wetlands and river networks). However, by checking the detailed information of these projects, including their locations, properties, and characteristics, we can conclude that, excluding the LID practices of municipal road sections, most other projects have already incorporated the elements of sponge city in their original plan. It is confusing as to whether the brand-new Sponge City Initiative can bring further improvements to urban resilience, and whether the funding is used appropriately. Therefore, it is difficult to evaluate and monitor the projects' effectiveness.

From a spatial distribution perspective, spatial distribution characteristics are based on Table 7.6. We see that most projects are located outside our research area (Gongming), and are in or around Fenghuangcheng. Gongming is more densely populated within Guangming New District, and is more vulnerable to urban flooding (Table 7.7 and Fig. 7.2). Nonetheless, there are only eight projects for building sponge bodies in Gongming, accounting for only 15.7% of the total number of projects. Among these eight projects, four cover the sewage systems (main road sections), three are rectification of river reaches, and one is the construction of an industrial park.

Although the drainage systems on roads and river systems for rainwater absorption are important, the sewage systems and physical environment of living communities play an important role in improving individual resilience. According to special planning for sponge city construction in Shenzhen, the core part of sponge city construction is the built-up urban areas with low ecological sensitivity to sponge city construction, which accounts for 29% of the total area of Shenzhen city. The quantity and scale of green lands in these areas are very few, and urban renewal should be actively concerted to sponge city construction" (p. 74).

However, in fact only one project planned for sponge city construction is urban redevelopment in nature, located outside Gongming. Based on the redevelopment planning of Shenzhen,[6] we find 13 urban redevelopment projects planned in Guangming New District in the upcoming years (2016–2020), and six located in our research area (Gongming) (Fig. 7.3). Nonetheless, in Gongming, none of the projects constitute a redevelopment project with elements of a sponge city.

Based on above-mentioned information, we can conclude that most projects reflect the mindset of "doing for the sake of doing", adding the so-called sponge bodies into existing projects so as to meet the central government's target. In addition, there are

[5] Source: Shenzhen government.

[6] http://www.szpl.gov.cn/xxgk/tzgg/csgxgg/

Table 7.6 Spatial distributions of planned sponge city projects

Project classification	Number of the projects	No. of projects distributed in Fenghuangcheng	No. of projects distributed in Gongming
Municipal roads construction	16	16	0
Construction zones (industrial park or communities)	10 (1 community)	9 (1 community)	1 (0 community)
Public park and green lands construction	5	5	0
Civil infrastructure construction (sewage systems)	11	7	4
Disaster prevention and water ecological restoration projects (wetlands and river networks)	9 (including 2 monitoring and research platforms)	4	3

Source Author summarized based on the project library on the special planning for sponge city construction

Table 7.7 Statistical data on urban-flooding-prone areas in 2014

Area	Number of urban flooding	Flooded areas (10,000 m^2)		Maximum depth of urban flooding (m)	Maximum flooded hours (h)	People displaced after disaster
		Total area	Maximum area			
Gongming	16	3.18	0.5	0.6	4	38
Guangming New District	26	4.48	0.5	0.8	4	42

Source Guangming Wind, Flood, and Drought Prevention Office

51 sponge city construction projects to be completed, but only eight are being carried out in Gongming, an area more vulnerable to urban flooding. Instead, most sponge bodies are being constructed in or around Fenghuangcheng, an area with better natural attributes to withstand urban flooding. Although there are six redevelopment projects located in Gongming, none of them involve sponge bodies.

As discussed before, sponge city is a presumably physical-environment-beneficial initiative that improves resilience. The special planning for sponge city construction in Shenzhen also stated that Shenzhen should make full use of urban redevelopment projects to realize the vision of a sponge city. Thus, it is worthy for us to understand why sponge bodies are seldom planned in Gongming nor added into redevelopment projects.

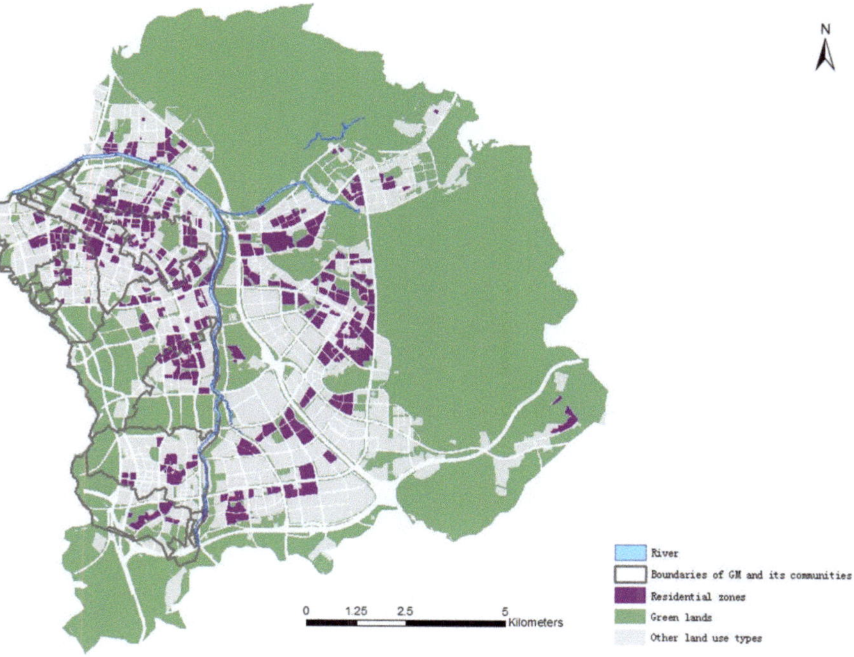

Fig. 7.2 Spatial distribution of residential zones in Guangming New District. *Source* Author

7.4.2.3 Likely Reasons for the Site Selection of Sponge Bodies

(1) Planning mechanism and spatial distribution of sponge bodies

Difficulty in coordinating similar plans and cooperation between government agencies restricts site selection for sponge bodies. Special planning for sponge cities often overlaps with drainage planning for stormwater, water security planning, green system planning, planning for rainwater drainage and waterlogging prevention. It is then necessary to combine and connect sponge city planning to these planning and specify the responsibility of different agencies. Planning coordination and cooperation between players and different departments costs time and money, reducing the efficiency of construction work and reducing the efficiency of construction work and complicating the site selection process and planning implementation for sponge city development. Planner D remarked:

> Sewage system construction and greening are controlled by the city administration, the mission of water quality belongs to the environmental protection department, dams are constructed by the water authority, and sponge city construction has a special and single office in Shenzhen Combining these similar work together needs cooperation between different departments. (Planner D)

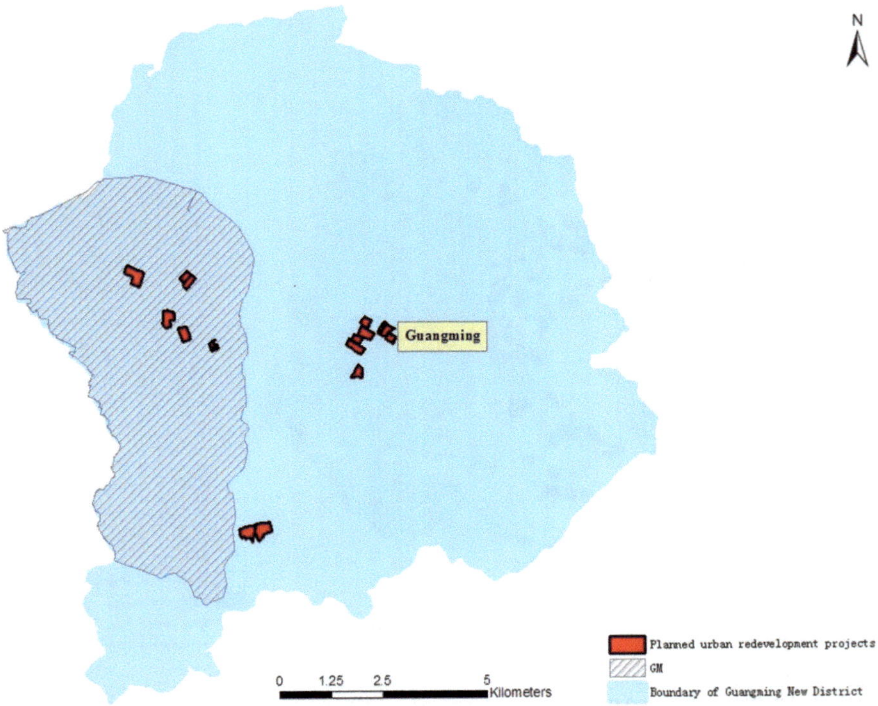

Fig. 7.3 Spatial distribution of planned urban redevelopment projects in Guangming New District. *Source* Author

As Planners C and A stated:

> Although expensive and accurate models can be applied to simulate urban flooding, the landscaping and engineering features in reality could be far more complicated. Applying the simulation results in the site selection process could be a huge challenge. (Planner C) City landscape and green land systems are large concepts, but the site selection for sponge bodies should be site-specific as their buildability in the site's context is key. Therefore, there is a pressing need for more cooperation between different departments. (Planner A)

Sponge city construction should spatially and temporally link these plans and be an integral part of other projects such as urban renewal, new town construction, and road construction. Therefore, site selections of sponge bodies are fettered by these constructions.

(2) Technology and spatial distribution of sponge bodies

The technology of sponge city construction is limited in China compared to elsewhere. At present, this type of construction is prohibitively expensive. Most construction is in the form of green infrastructure without costly permeable pavements. Therefore, areas with rich natural resources and landscapes are prioritized in serving as sponge bodies that will help reduce the cost and can be completed easily.

> Vertical greening is a high technology controlled by foreign developers. (Developer A)
>
> There is no technical guiding for sponge city construction, which increases the difficulty in converting the concept into reality. That is why sponge city construction in China is mainly concentrated in places with better natural resources. (Planner A)
>
> The technology of green building in foreign countries is more mature, which can result in low cost of construction and further help in saving electricity and energy to make profits in the long term. (Planner A)

(3) Financing mechanism and spatial distribution of sponge bodies

Sponge city construction is financed by the government funding for pilot projects, which only make use of the massive funding from central government to appeal to developers for sponge city construction. Considering the financing mechanism, PPP (Public–Private-Participation) for sponge city construction is still not mature enough to create market incentives in sponge city construction in the short term. PPP is based on a win–win desire among participants; however, it seems only the residents benefit in the short term, and they do not want to invest directly in this type of development.

Developers are at the forefront of the real estate market, and they are knowledgeable about the market. As developers, though the sponge city can bring profits in the long-term, the incentive of investing in non-profit projects like sponge body construction in the short term is weak. In addition, local governments are passive in this kind of development as these projects are less visible and less able to drive GDP growth.

Under these conditions, there are three results from forced sponge body construction. First, profitable developments, such as property-led redevelopments that can bring developers huge periodic profits, will stimulate the developers to bid for development rights from government. They even offer to build a sponge body to gain favor with governments to win the development rights. A developer remarked:

> The sponge body construction and design in redevelopment is always played as a gimmick to compete with others to win the bidding of the redevelopment project. (Developer B)

Governments are also willing to see the results, and they not only benefit from the land transfer, but also from transferring responsibility of the sponge body construction to developers. Even if there is not enough financial support for sponge body construction, the huge cost of sponge body construction is passed on to residents. In summary, it is a form of rights transfer from the municipal government to the developer under bounded profits, and residents pay for sponge city construction. It should be mentioned that those who can afford these properties are not always the existing residents. Those with low social resilience, low income, and low education cannot afford or rent these properties.

Second, new developments cannot generate large profits as the previous type of development (e.g., bundled with the construction of "affordable housing,") or are faced with more difficulties (e.g., involving houses with complex ownership). In this situation, whether the development can be carried out depends on the attitudes of developers. Even if there is funding support for sponge body construction, the huge cost and low periodic interests from the development leads to sponge bodies being low quality or constructed in places with better natural resources so as to save money.

To achieve the vision of a sponge city, governments tend to relax the regulatory requirements on its quality. It is a form of rights transfer from the municipal government to the developer without huge bounded profits, compromising however the effectiveness and quality of sponge bodies.

Based on these analyses, we can understand why the sponge city is being constructed in Fenghuangcheng instead of places with high population density and serious urban flooding problems. Integrated planning forces them to add sponge body construction to existing similar plans. If there is problem with site selection in similar plans, this will be enlarged during construction. Besides, technological limitations lead to huge costs in this type of construction and cause the projects to be clustered in places with better natural resources. Sponge city construction puts more effort in natural resource protection and improvement, ignoring the newly constructed projects, especially in communities under redevelopment. Moreover, considering its financing mechanism, it is difficult to persuade developers to build sponge bodies, especially in property-led redevelopment projects involving complex ownership that cannot generate considerable profits. Even if the developers are forced to build sponge bodies, the quality and post-maintenance still cannot be guaranteed. Developers instead offer to build sponge bodies for profitable development projects, which gives them more credence to obtain the development right of the present and future projects.

During these processes, governments seem to focus more on financing and the capacity (present planning and technologies) to carry out sponge body construction, ignoring or disregarding quality and effectiveness. The short-term visible results are the key driver for governments to pursue sponge city construction, which can be a plus to the officials' political achievement.

7.4.2.4 Summary

In summary, urban planning under the state-centric model shows that participation is always low and planning is always carried out according to the will of the state. The rationale for the government's decision on the site selection for sponge bodies is to maximize its political achievement. They want to choose places with good ecological and natural resources, which reduce cost and technical requirements and help them attain visible results within minimal time. It may be the case that the central government pushes Shenz hen to be a pilot city in sponge city construction to achieve demonstration effects for other cities with similar problems. The fast-food-style development, however, focuses more on the political aspect and ignores the inherent problems of the physical environment embedded in the drainage capacity of old urban areas, meaning that the initiative is unlikely to solve present problems. This does not mean there is no point in constructing a sponge city—in the long-term, if this type of construction is required for all the developments of the city, it would be able to improve water ecology and facilitate sustainable water circulation. The present development, however, cannot improve the poor physical environment due to the rather unsystematic site selection of sponge bodies, ignoring old and poor communities such as those in Gongming.

We can conclude that the resilience-oriented measure of Sponge City Initiative cannot nurture individual resilience for the residents of Gongming by improving the physical environment of poor communities. For the reasons mentioned above, the government restricts the site selection for sponge bodies that can improve the physical environment of communities.

7.5 Place-Responsive Measures for Individual Resilience Nurturing

Since the initiative on the national scale (sponge city) cannot nurture individual resilience, in order to improve individual resilience, as pointed out in chapter two, place-responsive measures are the best choice to realize DRR. As noted in Chap. 6, community management embedded into housing/property management can change individual resilience performance. Personal perceptions are closely related to the changes of the physical environment and community management, which also suggest that the improvement of the physical environment and community management can enhance individual resilience performance. To conclude, urban management can influence what the residents think about disasters and how they find solutions to cope with and adapt to disasters.

Although the literature pointed out that urban management against disasters should be localized, the place-responsive concept has not received much attention in the management of urban flooding (Cash et al., 2003; Cutter & Finch, 2008; Cutter, 2012). The measures should depend on the context of the community and individual situations. Therefore, this research describes place-responsive measures to manage the urban flooding regarding vulnerable residents' individual resilience performance in Gongming. The previous interaction analyses among people, disaster, physical environment and social environment provide the foundation for the formulation of measures. Rethinking urban management for urban flooding mitigation by integrating with resilience theory has the following implications:

(1) Social aspects, instead of physical approaches, should be focused on in mitigating natural disturbances (Maclean et al., 2017; Stone-Jovicich et al., 2018).
(2) The nurturing of individual resilience should be part of ongoing policies and programs (Vedung, 2017).
    a. Mitigation measures should emphasize the improvement of residents' disaster awareness.
    b. Mitigation measures should emphasize the internal structures of communities in society rather than external forces.

c. Mitigation measures should pay attention to the physical environment of communities, which is closely related to the perceptions of residents and human behavior.

(3) Policies, laws and programs should be intervened to develop social learning paths of residents, which will fundamentally change their behavior (Ison et al., 2007).

7.5.1 Intervention on Interactions Between People and Disaster: Disaster Awareness Improvement

The analyses on the interactions between people and disaster show that the ones who suffered the worst urban flooding are more likely to have high disaster awareness and consequently more prepared actions against urban flooding.

Therefore, the following measures can be adopted to improve individual resilience performance. First, education can improve resilience (Sugimoto et al., 2010; Tanaka, 2005). The literature suggests that information processing consists of stages: hearing the information, understanding it and then perceiving its relevance. Education is the most direct way for individuals to learn and shape the degree to which they assess and perceive the risks (Amri et al., 2016; Tanaka, 2005). Education enables residents to prepare for the disasters by showing resilience-oriented behavior. Therefore, there should be education on urban flooding on residential scales. Based on the regression analysis, education level as a control variable does not have significant influence on changes of performance of individual resilience. Although the role of formal education on disaster preparedness is very important, on a place-responsive basis, informal education seems to be able to stimulate and intensify the awareness of disasters while formal education cannot.

Based on the modern experiences from Japan in earthquakes, we see that disaster drills or exercises can help residents to acquire the measures for coping with and adapting to urban flooding. The drills also can help residents to improve their disaster awareness by understanding the dangers of disasters. Therefore, it is necessary to encourage and require the community and property manager to drill the residents on coping with urban floods.

It is also necessary to make full use of residents' demands for public participation in planning and management. The survey data shows that the demand of public participation on management and planning is high. But around 55% of the surveyed residents do not know how to participate in. No more than 1% of them have been invited to participate in urban flooding management. Therefore, the community managers should encourage organizations to meet the demands of residents to participate in urban flooding management.

7.5.2 Intervention on Interactions Between People and Community Management: Internal Structures in Societal Improvement

The internal structure of society is very important for individual resilience improvement (Rigg & Oven, 2015). On the one hand, people are responsible to their own management, which may influence their resilience. On the other hand, it is not difficult to find that the measures for disaster awareness improvement based on a better social environment, such as information sharing, disaster drills and informal education, all call for the healthy social climate of the communities (Rigg & Oven, 2015; Welsh, 2014). On a place-responsive basis, mutual trust and perception of life in one's residential area are two important factors influencing individual resilience performance of the study area.

Well being as perceived by residents and mutual trust are important components in building a social environment in a community (Guo et al., 2014; Sampson, 2003). Denser social ties help to exchange information and mutual assistance, which can be readily perceived by residents, especially the ones whose activities are within the communities (e.g. stay-at-home mothers and the elderly). Research suggests that females are more likely to have denser social ties that connect them with more neighbors than male ones (Belle et al., 2014; Curran & Saguy, 2001; Mills, 1997). Therefore, they are better catalysts for social environment building. However, women have less access to resources and to information channels. According to previous analyses, coping and adaptation measures should be tailored for peasant workers. To improve individual resilience performance, on a place-responsive basis, the main focus of the external intervention should improve social environment building on a scale of residential zone, especially for the female and peasant workers living in urban villages. Property managers or community leaders should provide cognitive support to the residents, including the provision of diverse information, new knowledge and advice, which can activate the self-learning of residents to urban flooding through the changes of their perceptions of their management, and further simulate the human behavior towards resilience-oriented. Besides, the provision of tangible aid and services such as support and childcare on a scale of residential zone can contribute to individual resilience. Grassroots groups, such as NGOs must be invited to improve the social environment (Islam & Walkerden, 2015; Ramalingam, 2015), especially when communities have a large proportion of peasant workers without strong attachment to the community.

7.5.3 Intervention on Interactions Between People and Physical Environment: Physical Environment Improvement

The living environment of residents and waste management of communities play a dominant role in changing individual resilience performance. The interaction theory between human behavior and physical environment proves that the changes of the physical environment of communities can influence human behavior (Stokols, 1978; Stokols & Altman, 1987; Stokols et al., 2013). Changes and improvement of the physical environment depend on the different levels of governments as well as property managers or managers of urban villages if they are not commercial properties. Therefore, the efficient intervention for individual resilience improvement of residents depends on the government learning on the present situations and attempts to improve the present physical environment.

In the study area, 80% of the communities are urban villages without specialized managers to manage them. The infrastructures, especially the sewage systems are deficient. All these induce poor environment and inefficient waste management. On a place-responsive basis, external intervention should focus on waste management. Waste management does not only influence personal perceptions and lead to different behavior of residents, but also have significant impacts on human health. The pace of urban renewal and improvement of urban villages should be accelerated. As mentioned above, there are many problems related to land ownerships left over from the past, which led to lots of urban villages in the study area. Although it is difficult, it still requires collaborative governance to solve the problems as soon as possible.

In summary, local interactions among people, disaster, place and society suggest that the place-responsive arrangements should consist of the following. Informal education and natural disaster drills should be developed on a community scale for disaster awareness improvement. Property managers or the urban village managers should emphasize the building of social environment in their communities through the information sharing and social services provision. Finally, physical environment including waste management and the present living environment, especially for the urban villages, should be improved based on the efforts of collaborative government.

7.6 Discussion and Conclusion

This research chooses Gongming as a case for exploring the potential for resilience nurturing under the present management regimes. Literature focuses on using theoretical and empirical examples to support management that can facilitate the building of cities, which are resilient to unpredictable changes. There are few empirical cases that explore the degree to which resilience depends on the urban management. This chapter therefore draws on the social learning theory to examine and explain the issues related to self-learning, collective learning, and government learning for the residents of Gongming on resilience nurturing, based on the identification of the People-Disaster-Place-Society associations that have limited individual resilience performance. Finally, based on the previous research, place-responsive arrangements are identified, which help increase the individual resilience of the residents of Gongming. Based on this research, there are four groups of findings, which can be further discussed.

For interactions between people and natural disturbances, low disaster awareness is found to be the dominant reason for poor self-learning. People who exhibit low disaster awareness are always those with low income and limited education and without an understanding of disaster prevention. These groups of people form a new type of imbalance in information flow, which reduces social resilience and increases the risk of exposure to urban flooding. It is therefore necessary for governments to nurture their social resilience. However, governments lack the consciousness or desire to do so. The good news is that Tencent wants to take responsibility for public education and information dissemination on disaster awareness improvement, as well as the early-warnings on extreme weather through the Wechat platform. It is helpful for most residents; however, it would still not be tailored to the specific population who may not be able to obtain this type of information. In developed countries, extremely vulnerable populations with low individual resilience are closely monitored. Informational and structural measures could help them disseminate information to everyone, but the measures are further tailored to serve different ethnic groups in certain geographical areas. Compared to the advanced strategies, as well as problems faced in reality, we can conclude that the resilience-oriented measure of the sponge city initiative does not have the necessary conditions for facilitating self-learning to improve individual resilience.

For interactions between people and management, the poor social environment between individuals across different scales and levels block the local groups' ability to self-organize, learn collectively, and adapt to and shape changes. This is particularly evident in the communities with many low-income peasant workers, as these communities tend to be characterized by poor living environment, poor housing management, and substandard infrastructure. Formal grassroots managements get too much work from higher governments. Heavy workload makes them focus more on administrative work, and less on providing social services such as disaster education. At the same time, informal grassroots organizations, such as the relationships of clan and blood, or NGOs, also cannot provide these services. It is difficult for the non-local residents to blend into clan organizations. The NGOs in China still tend to serve the government and no NGOs offer social services to the extremely vulnerable populations with low individual resilience. Facing this situation, the initiative of constructing a sponge city has attempted to improve it, but the policy has not focused on nurturing resilience in the social dimension. The NGOs participating in the sponge city construction claim that governments put more efforts in infrastructure construction than social resilience nurturing. NGOs and some other social associations focus on the post-disaster relief work, ignoring education and information dissemination before disasters. Compared to western countries, NGOs in Shenzhen, which serve the government in sponge city construction, focus more on making suggestions on management modes for the municipal government and exploring financing modes for sponge city construction than on providing social services to poor residents (Islam & Walkerden, 2015; Ramalingam, 2015). Although they are important strategies for sponge city construction, poorly perceived social environment are the key constraint in the improvement of individual resilience.

The limiting factor for individual resilience performance suggests that urbanized, poor areas should increase physical environment such as the sewage systems, particularly ecological infrastructure. Since sponge city construction focuses on physical infrastructure construction, it is recommended that the physical environment of Gongming should be improved. It was identified however that the Gongming's physical environment could not be improved in a short term. Based on the special urban planning for sponge city construction of Shenzhen we found that the areas that need the sponge bodies to improve physical environment will not receive these types of infrastructures right away. Areas that have a better physical environment are always selected for the construction of sponge bodies. There are several reasons for this. First, sponge city construction not only solves the problems of urban flooding, but it also purifies rainwater. Places with green lands have water systems that can undertake such work. The second reasons is that planning restricts the construction of sponge bodies, which should be embedded into planning activities. Better ecological resources will help to reduce the cost of this type of construction. Finally, most projects are surrounding a new town named Fenghuangcheng, which is a new core development area of the Guangming New District, and it is the strategic growth area for Guangming. It can be seen that rich natural resources, good physical environment and profitable development here facilitate sponge body construction. However, these places have low population density with few urban flooding. Considering the limited improvements to the physical environment for the vulnerable areas, the resilience-oriented measure fails to realize the goal of nurturing individual resilience in the short term.

Based on the analyses, we can conclude that the resilience-oriented measure, based on government learning, has still not made up for the deficiencies of self-learning, collective learning of residents, and their previous failure in learning for resilience nurturing. As the national measure for resilience nurturing, Sponge City Initiative was identified that it cannot perform the functions as the ones used and tested in developed countries. The social learning of the government is still on the second loop without any advancement towards the third loop, limiting the government in nurturing resilience, especially in social resilience nurturing. There are three evident incentives facilitating this. The first incentive is political achievements. Social resilience is a process that takes time and money, and it is difficult to quantify and evaluate the political achievements in nurturing social resilience. However, a tangible infrastructure can be seen and observed. As the central government assigns the construction targets to pilot cities, the problem of focusing too much on physical construction instead of social resilience nurturing is a natural result. The political achievements pursued mismatch the ones the government should do, which led to the dislocation of sponge bodies, resulting in failure of physical environment improvement for the vulnerable areas. The second incentive is the market. Nurturing resilience should be a type of social service provided by the government. However, the state, as the main non-profit service sector, has different pressures, especially in financing. Therefore, local governments must seek cooperation with the private sector for funding. As the residents who benefit from the projects do not want to pay, the government must resort to private developers. Nonetheless, as development is a market-oriented activity, developers are only willing to pay for the construction if the project is profit-making. In this case, the developers would offer to build sponge bodies to please the government to lure for development rights, but meanwhile this also means that the government is transferring the responsibility of sponge body construction to the developers. Finally, those who bought or rent the buildings in newly developed regions will pay for sponge city construction. However, people who can afford homes in the new town are not those with low resilience. This mismatch of the market leads to poor maintenance in sponge city construction. The third incentive is service demands of residents from governments on urban flooding mitigation. Urban flooding is dangerous, particularly for children and the elderly. Disease and contaminant transmission through rainwater may also threaten residents' health. However, the residents do not notice it because of their low disaster awareness. Perceived demands and actual demands for resilience nurturing are dislocated. This dislocation of resilience nurturing demands leads to the final failure of resilience improvement.

Social learning occurs in social environment between individuals and within social groups, which can be influenced by the intervention of governments (Henly-Shepard et al., 2015; Smith et al., 2016). As the sponge city initiative cannot nurture individual resilience, measures on a place-responsive basis, as derived from previous research, are required. More specifically, education, especially informal education and natural disaster drills should be conducted on a community scale for disaster awareness improvement. Social environment building is necessary for individual resilience improvement. It requires the property managers or the urban village managers to improve information-sharing paths and enhance social service quality. It needs to be mentioned that female play a dominant role in information sharing, and they have the potential to participate in the urban flooding management of their communities. Last but not the least, as mentioned above, there are complicated problems of urban villages have not been solved until today, which leads to the poor living environment of urban villages cannot be improved in a short term. Results show that personal perceptions of the physical environment play a dominant role in simulating the resilience behavior of individuals. More specifically, waste management and the present living environment for the urban villages should be improved based on the efforts of collaborative government.

Chapter 8
Conclusion

8.1 Introduction

This research constructs a theoretical framework and uses an empirical case to explore the potential for nurturing urban resilience under the current urban management based on individual resilience evaluation and its limiting factors identification. Current performance of individual resilience is framed into the capacities of coping and adaptation. Personal perceptions of urban flooding, the physical environment and social environment are focused on for they influence individual resilience performance based on the human–environment interaction theory. Urban management is assumed a political intervention, which can enhance individual resilience performance by changing residents' behavior to a resilience-oriented manner.

There is a logical relation behind the analyses that social learning can change residents' behavior to be resilience-oriented, considering the self-learning and collective learning are needed to create the positive feedback loops to lead to resilience. However, in reality, the effects of self-learning and collective learning are not always significant, which is why local residents' resilience are not always high. At this time, governments' measures can contribute to the changes in behavior of individuals and communities to become more resilience-oriented. Since the processes of social learning occur in the society of individuals and social groups under the present urban management, and it is therefore necessary to check if the strategic measure currently can nurture individual resilience. Based on this logical relationship, an empirical case at the sub-district scale in Shenzhen was selected to evaluate individual resilience and identify the factors of personal perceptions limiting residents' resilience performance. Limiting factors can reflect the effects of social learning, and identify reasons for the failure of social learning. Based on these analyses, to identify the effects of institutional measures in nurturing individual resilience, the Sponge City Initiative was selected as a resilience-oriented measure in this research to determine if the intervention can nurture individual resilience by considering its effects on activating the self-learning and collective learning of local residents, especially those with low

J. Song, *Individual Resilience to Urban Flooding and the Implications for Urban Management*, https://doi.org/10.1007/978-981-97-0546-7_8

individual resilience. Based on the previous analyses, place-responsive measures were provided for nurturing individual resilience for residents of Gongming.

Quantitative and qualitative methods were used to conduct the research. Individual resilience performance was evaluated based on individuals' self-reported behavior. Factors influencing individual resilience performance were identified from questionnaires and interviews. Furthermore, HLM was applied to identify factors influencing and limiting individual resilience performance. Following this, social learning theory was applied to examine and explore the problems of self-learning and collective learning towards nurturing resilience based on the identified factors limiting resilience performance. Finally, the institutional measure, represented by the Sponge City Initiative, and its ability to nurture and make up the deficiency of social learning is analyzed. Based on previous analyses, place-responsive measures were summarized for individual resilience improvement. Then the findings are summarized and discussions are drawn.

8.2 Major Findings and Discussions

The first group of findings answered the first group of research questions: "What was the current performance of individual resilience of residents vulnerable to urban flooding? And did the individual resilience vary among different communities?".

8.2.1 Strategies Adopted by Individuals and Their Communities

Many of the surveyed residents choose the following strategies, like "disseminate safe-haven news to your families and friends after you have known it," "do some preparation for heavy rain or urban flooding," "have the habit to check weather report" and "follow the early-warning signal to strictly arrange your travel." These strategies are the ones that depend largely on information sharing and communication. Therefore, it seems that people in Gongming like to share and communicate information if they have known it. Besides these strategies, there were some that are seldom adopted by residents in Gongming, including "know the early-warning signal of heavy rain," "know which areas usually experience urban flooding around my living communities," "My living community has some innovative strategies for adapting to disasters," "My living community provides education for improving your capacity for adapting to disasters," "My living community has the daily maintenance about the pipelines and pipe junctions" and "My living community has some measurements to overcome the urban flooding." The group of strategies adopted by a moderate number of residents contain "have learnt how to help myself escape from disasters," "know where I can escape from the heavy rain or urban flooding when I am outside," "take the initiative to study the knowledge of disaster prevention and risk aversion," "I am willing to participate into the new strategies building for disaster prevention of my living community" and "my living community has commissioners to repair the pipelines after they have been jammed." Comparing these three groups, we can find that usually the residents of Gongming show the willingness and confidence to overcome the urban flooding. Nonetheless, they in fact have not acquired the necessary knowledge, methods and strategies to cope with and adapt to urban flooding.

8.2.2 *Current Performance of Individual Resilience of Gongming*

The mean score of individual resilience performance is 4.62, falling under the intermediate level. As mentioned in the research methods explained Chap. 4, combining the survey data of residents of Gongming, the evaluation of individual resilience is on a relative basis, and even the ones with high mark of their resilience performance may not behave with the desired level of resilience, let alone that most residents in Gongming show only intermediate level of individual resilience performance. The calculated individual resilience performance suggests that residents of Gongming did not have enough resilience to reduce the risk of being extremely vulnerable to urban flooding. In one sense, it means the groups for which resilience improvement measures should be tailored are larger in Gongming. In addition, it suggests that there is still room for the residents of Gongming to improve their resilience performance.

8.2.3 *Variations of Current Performance of Individual Resilience in Different Communities*

Results show that the variations of individual resilience reflect spatial discrepancy under three groups. Residents of the communities located in the middle part of Gongming are more likely to have higher individual resilience than residents of other areas. Residents of communities in the northwestern and southern parts of Gongming are more likely to have low individual resilience. The individual resilience of the residents residing in the northern part of Gongming is between that of the other two spatial groups. It on the one hand suggests that community attributes can influence individual resilience, and on the other reflects that there should be some common attributes for the individuals who live in similar communities.

The second group of findings answered the second research question: "What were the determinants (People-Disaster-Place-Society) of individual resilience?".

8.2.4 Factors Influencing Individual Resilience Performance

Interactions among people, disaster, place and society were analyzed in detail in this study, and these factors are embedded into the factors showing personal perceptions of urban flooding, the physical environment and the social environment of their communities. HLM was conducted to identify the factors limiting the performance of individual resilience at the individual level while considering the influences of attributes at the community level.

The findings of this study suggest that the common personal attributes and disaster awareness distinguishing low and high individual resilience were as follows. People with low individual resilience are more likely to lack the attributes to offer leadership in times of difficulties. They did not want to participate in urban planning and management activities related to urban flooding. Residents who may not care about susceptible groups such as children are more likely to have low individual resilience. Furthermore, residents were more likely to have low individual resilience if they have not had the experience of losing their possessions in an urban flooding. Similarly, people who have not received help or support from governments and their workplaces to cope with urban flooding do not seem to have the high individual resilience that recipients of aid did.

The findings of this study suggest the common attributes of physical environment of communities that distinguish low and high individual resilience. Although the physical environment could not distinguish people from different performances of individual resilience, the perceptions of the quality of physical environment, including the explanatory variables of the living environment of communities and waste management of communities, are closely related to the changes of individual resilience performance. Based on the regression results, it can be concluded that the residents who feel bad about their physical environment are more likely to have low individual resilience. Besides, the higher residents' resilience, the better they feel about their waste management. Therefore, we can conclude that a clean and high quality of physical and ecological environment can increase the place attachment and the resilience-oriented behavior of residents. The better and cleaner the physical environment, the more likely the residents are to show high individual resilience.

The findings of this study also suggest some common personal attributes that distinguish low and high individual resilience in terms of different social environment of communities. Based on the results of the regression, it can be concluded that both social environment as well as the perceptions of social environment can significantly influence individual resilience. At the community level, gated communities with better property management have a positive role in individual resilience improvement. It suggests that no matter the property is commercial or non-commercial, if they are gated communities with professional managers, the management can facilitate residents to build individual resilience. It also suggests that community management can affect residents' behavior, and proves that place-responsive community management can increase individual resilience. At the individual level, people with high individual resilience demonstrate better social environment within their communities, including the explanatory factor of social cohesion. More specifically, those with low level of individual resilience are more likely to demonstrate common attributes such as weak social cohesion with no feeling of well-being about life in their communities, little information sharing and few public activities among residents.

To conclude, the major limiting factors for the individual resilience performance of Gongming encompass low personal disaster awareness, poor social environment in communities and poor quality of built and physical environment perceived by residents.

The third group of findings answered the third group of research questions: "Whether the strategic measure of Sponge City Initiative could help nurture individual resilience? And what were the place-responsive measures for individual resilience nurturing?".

8.2.5 Analyses and the Role of Institutional Measures in Social Learning to Nurture Resilience

One attribute of residents with low individual resilience in Gongming is their low disaster awareness. Based on the social learning theory, it can be concluded that problems exist in the self-learning process in nurturing individual resilience.

The analysis of residents' low disaster awareness is based on the correlation analyses of the limiting factors, which consist of disaster awareness and resilience-related personal characteristics. Responses to the questionnaires and interviews supported these results, indicating that the social group, which is young, single, low-income, male, and living in houses with limited rights is more likely to demonstrate low disaster awareness. This forms a new type of imbalance in information flow.

Therefore, it is hoped that Sponge City Initiative, as a resilience-oriented measure, would reduce the imbalance in information flow to improve disaster awareness. Embedded in the place-responsive measures of sponge city projects, the pilot LID projects which could supposedly improve disaster awareness, and the Smart Sponge City projects which can supposedly disseminate information to residents to improve disaster awareness, are failing. This failure can be attributed to these measures not being tailored to specific population groups, in turn exacerbating imbalance in information flow. The aims of the Sponge City Initiative, as stated in the news and special planning, are to increase awareness of disaster prevention and to conserve water for environmental protection. However, only the planners, engineers, and other local officials have visited the completed LID projects, reflecting that the projects are not focused on nurturing residents' individual resilience. The WeChat platform from Tencent can spread information on disaster awareness to most residents. However, the platform is not tailored to the specific population group with low individual resilience, as they may not be willing to accept information on dealing with urban flooding.

In summary, the Sponge City Initiative aims to improve residents' awareness of disasters and environmental protection. However, its implementation indicates that the government focuses more on the technologies and speed of sponge body construction, ignoring the nurturing of residents' individual resilience, especially of those who have trouble obtaining information.

Another explanatory attribute for the low individual resilience in Gongming is poor social environment *led by the poor grassroots management.* Based on social learning theory, it can be concluded that there are problems in collective learning regarding the nurturing of individual resilience.

The residents' poor social environment can be diagnosed and deduced based on the interview responses and the land use and statutory plan maps of Gongming. Spatial data shows that rural residence land (urban villages) accounts for 80.11% of all residential land. Most rural communities do not have property management, which increases the difficulties experienced by the communal committee in managing residents. There are additional findings from the interviews with staff in the community committee and neighborhood committee as well as local residents. For formal grassroots management, the upper government assigns them too much administrative work. In addition, the staffs are of low quality and their projects underfunded. The rights and obligations of the sub-district and communal committees are usually ambiguous. These factors mean that formal grassroots management focuses more on administrative work at the expense of collective learning, ignoring social services such as public education and information sharing. For informal grassroots management, the RSC includes clay and blood relationships, and seems to play a more important role in motivating collective action. However, these collective actions push aside the non-native population, limiting the efficiency of collective learning.

Therefore, it is hoped that sponge city construction as a resilience-oriented measure can improve the social environment needed to realize collective learning. However, the aspect is not explicitly mentioned despite the existence of special plans for sponge city development. Staff of grassroots management organizations have confirmed that until now, they have not participated in or been assigned tasks related to sponge city development for building social environment. Under this situation, the improvement of social environment relies largely on informal grassroots social groups. Outside of China, NGOs have a dominant role in social environment building. Therefore, embedded in the place-responsive measures of the sponge city development projects, NGOs' roles in improving social environment were examined. NGOs were invited to participate in sponge city development. Outside of China, NGOs emphasize social services such as public education for extremely vulnerable people, which may improve social environment, especially for those with low individual resilience. However, the NGOs participating in the sponge city development in Shenzhen focused on governmental services such as exploring financing models and government management services, as their participation in governmental work determines their source of funding and power under the state-centric model.

In summary, the resilience-oriented measure of sponge city construction does not demonstrate attempts to build social environment. Social environment is the primary constraint of resilience performance. Therefore, it can be concluded that

China's resilience-oriented measure ignores the nurturing of individual resilience. Furthermore, centralized governments do not provide grassroots social groups with enough power and freedom to play their roles such as providing social services for those with low resilience, causing their difficulties in realizing collective learning.

Another attribute of residents regarding the low individual resilience in Gongming is the poor quality of physical environment. Low quality of waste management and living environment, as well as an inefficient sewage system, prove that government learning on sewage system management for Gongming has failed.

The causes of the failure of government learning can be explained based on the interview responses and secondary data from websites. As mentioned, too much rural residence land (urban villages) developed by underground developers has underdeveloped and insufficient infrastructure, including the sewage system. Moreover, the extreme weather conditions caused by climate change have increased the pressure on governments to implement methods to cope with and adapt to extreme weather. The growth in and pursuit for GDP make the government ignore the significance of ecological land in absorbing and purifying rainwater. In addition, the speed of installing sewage systems is slower than the rapid development of numerous areas. All these factors contribute to the failure of government learning to provide efficient sewage systems in Gongming.

Sponge City Initiative, as a resilience-oriented measure, should compensate for the deficiencies in government learning. Embedded in the place-responsive measures for sponge city development, the rationality and reasonableness for the site selection for sponge bodies, i.e. ecological sewerage systems, are analyzed. However, the findings suggest that these measures cannot improve the sewage systems to the required extent in the short term (2016–2020), according to the special planning for sponge city development.

To conclude, urban planning under the state-centric model is characterized by narrow participation and planning carried out based on the demands and willingness of governments. The young technologies used in sponge city development such as permeable pavements and green buildings are prohibitively expensive. Here, the main force driving governments to select locations for sponge bodies is the evaluation of their political achievements. They want to select locations with better ecological and natural resources to reduce costs and technical requirements so that their achievement can be manifested in a short period of time. Furthermore, pressure from the central government on sponge city development cause the governments of Shenzhen to emphasize the speed and pace of development, rather than the effectiveness of construction and post-construction maintenance. The high cost of construction in the short term blocks social financing, and developers pursuing periodic interest do not want to invest in this type of construction unless the project is profitable. Even if forced into the development, developers will choose cheaper measures, such as building green belt instead of expensive permeable pavements. To satisfy the targets of construction, governments relax inspection and approval standards for sponge body development. In this process, the primary aim of solving drainage problems is forgotten. Of course, this does not imply that sponge cities are useless—in the long term, if this type of structure is built throughout the city, water ecology must be improved and sustainable water circulation realized. However, this will not tackle the problems of poor quality of physical environment, especially in older and poorer communities such as those in Gongming.

Failure of government learning leads to this situation. If the government is unable to make the qualitative leap from the second to third loop of learning, it will inevitably overlook the nurturing of individual resilience. Therefore, it can be concluded that the resilience-oriented strategy cannot nurture individual resilience in Gongming.

8.2.6 Place-Responsive Measures for Individual Resilience Nurturing

In terms of the interactions between people and urban flooding, for Gongming, residents need drills under an urban flooding scenario. Considering the high demand of vulnerable residents to participate in the urban planning and management against urban flooding for their communities, education should be offered to residents. The lack of learning paths, instead of education, is the dominant limiting factor for individual resilience performance.

In terms of the interactions between people and social environment of Gongming, the poor social environment should be improved. Place-responsive measures include activating the roles of females, who usually are good at information sharing and more caring to the sensitive population such as children. Besides, urban management should institutionalize the provision of tailored assistance and services to special groups, such as single low-income migrants. They are overconfident or have struggles living in Shenzhen, with extremely low disaster awareness and low capacity to cope with urban flooding. Further, social and cultural development at the residential level is important to improve the social environment, such as the social services and help for childcare with the participation of NGOs. These can facilitate the residents of these communities behave in a more resilience-oriented manner.

In terms of the interactions between people and the physical environment, ecological infrastructure should be improved, especially for urban villages. Poor physical environment, including the dirty environment and irregular waste management can influence residents' perceptions, which in turn affect their behavior. Therefore, besides improving waste management for the communities, an urgent but difficult task is to deal with the problems of urban villages. The living and physical environments, including green space and basic infrastructure, of urban villages are poor. Nonetheless, if the improvement to urban villages leads to substantial increases in rents, new problems could arise as well. Seeking a balance would be key in the future.

8.3 Contributions

8.3.1 Theoretical Implications

The research on urban resilience in China is not as advanced as the USA, the UK, Japan and France. A few studies in China question whether resilience is amenable to urban management and regimes. Therefore, it is necessary to identify the interaction between urban resilience and management.

Considering the research scale, other countries have been more mindful of community resilience to urban floods than China. As the foundation of community resilience, individual resilience was selected as the research scale in this research. It is hoped that this will compensate for the limited theory on urban resilience to urban floods in China.

Moreover, the literature fails to identify the influencing factors and the dynamics between vulnerability, resilience, and management from the perspective of personal perceptions. Filling this gap is also a contribution of this research. The subjective and individual-centered personal perceptions to disasters, environment and management are self-reported by the surveyed residents and then applied to explore their influences on individual resilience. This methodology is another contribution of this research. In short, it is necessary to explore and quantitatively analyze individual resilience to urban flooding in sub-district undergoing rapid urbanization in China so as to complement the international research on resilience.

8.3.2 Empirical Implication

Even less empirical research on urban resilience has been done in China. To address the increasing severity of problems associated with urban flooding, the central government is applying resilience-oriented measures. However, the government has not enacted specific measures to create resilient cities; the Sponge City Initiative is closest of this kind in China. The Initiative was introduced in 2013, and Shenzhen had the winning bid for pilot sponge city development in 2016. As of the time of writing (summer 2017), the special planning for sponge city development in Shenzhen has been completed, as has the preparation for the first phase. Since 2010, LID practices have been tested in a small-scale development in Guangming New District, Shenzhen. These make it possible to explore the potential of a resilience-oriented measure in nurturing resilience. In this study, this evaluation was accomplished by using the case of Gongming. This helps to translate resilience into practice in China, and provides a reference for the state to better nurture residents' resilience, especially of those with low resilience.

On an international perspective, for the developed countries, this research develops a new pathway to explore the determinants of individual resilience. Although in developed countries, they begin to focus on the roles played by human perceptions and cognitions on their behaviors, it is seldom applied into the individual resilience nurturing. Until now, most of the existing literature only focuses on the influence of disaster awareness of human beings on human resilience, while the roles played by the perceptions of social environment and physical environment of human beings have been ignored.

For the developing countries, the empirical research contributes to them that, urban planning should put more emphasis on the perceived importance of individual resilience among these social groups. Such planning strategies, which affect and alter different social groups' perception of the physical and social environments, deserve further exploration in future research for urban managers, urban scholars and urban planners.

8.3.3 Policy Implication

Different from the literature in which policy suggestions are always at a city, regional, or even national scale, this research develops the place-responsive measures for individual resilience improvement, especially for the social groups residing in poor living zones. Although this operation may be difficult, it is evident that the high-order policy is incapable of addressing the increasing social inequality in a city under rapid urbanization. Therefore, it is necessary to offer suggestions, policies and measures for a small scale—here at the community scale—to better serve its residents.

In order to realize these policies, the implication for urban management is summarized as follows:

First, collaborative governance model instead of the state centric model should be adopted for urban flooding management and individual resilience nurturing. This model can facilitate cooperation between different departments of government, different types of urban planning, as well as different types of technologies from the perspective of institutional superstructure.

Second, Public–Private-Partnership (PPP) should be encouraged for individual resilience nurturing program. However, PPP needs to be improved which make it does not only pursuit the economic profits but also the social profits which can benefit to residents. Therefore, the public participation and demands of residents on public participation should be encouraged and valued by government.

Third, legal and political supports are necessary from a higher-level government, i.e. central government. In order to realize individual resilience nurturing, the legal supervision and monitoring on the efficiency of individual resilience nurturing program and the use of funds before, during and after the program are quite necessary.

Last but not the least, in the Chinese context, the central government has proposed several initiatives, most notably the Sponge City Initiative, to tackle the challenges posed by urban flooding. In order to improve individual resilience with due regard to its social dimension, thereby the Office of Sponge City in Shenzhen should optimize the Sponge City Initiative to add the social nurturing and improvement into the existing Sponge City Initiative.

8.4 Limitations and Recommendations for Future Research

8.4.1 Limitations Regarding Strategy Selection for Resilience Evaluation

In this research, individual resilience performance is evaluated based on the 15 selected coping and adaptation strategies adopted by the surveyed residents. Although the 15 strategies are the representative ones identified based on the literature review and interviews with scholars and governmental officials, this list is yet to be comprehensive enough. Some creative strategies have not been discussed, and further research should be conducted.

8.4.2 Limitations Regarding the Selection of the Factors Influencing Individual Resilience

Factors reflecting the social environment on the community level need to be explored in the future research. "Gatedness of the community" is proved to be more of a statistical significant factor to individual resilience. Therefore, the factors, which can better explain the social environment of living communities should be identified.

Factors reflecting the personal perceptions are the economic, social, cultural, and political aspects of the natural disturbances, geographic setting of places and social environment (Gaillard, 2007, Norris et al., 2008). However, these factors may vary at different scales of analysis. Therefore, not all of these factors are included in the scope of this research. Some factors may be more significant in the specific contexts of Gongming but less in other areas, vice versa.

8.4.3 Limitations of the Questionnaire and Interviews

The subjective questionnaire approach to measuring resilience to urban floods is used to reflect the actual capacity of individuals and communities to cope with and adapt to urban flooding. Resilience is a multidimensional concept, constructed by multiple factors in a localized context. However, the use of this subjective questionnaire and interview approach to obtain the perceived capacity at the individual level to cope with or adapt to urban flooding may be prone to some limitations in explaining the resilience of individuals to future urban flooding. This is a key limitation of this research.

8.4.4 Recommendation for Future Research

Two limitations need to be addressed in future research. Individual resilience is evaluated on a relative basis in this research. If this evaluation method can be widely accepted and emulated elsewhere, the list of strategies selected should be more comprehensive. The options settings for questionnaire should reference to individual health research,[1] and a standardized three-item five-point scale is more suitable and persuasive in measuring individual resilience. In addition, although this research set a scenario of urban flooding disaster for residents in answering the questionnaire, it lacks accurate descriptions of the scenario. It is important for future research to integrate changing scenarios into the questionnaires and interviews to assess the resilience of residents in the research area.

8.5 Conclusion

This chapter summarizes the major findings and conclusions of the research on individual resilience to urban flooding, including individual resilience evaluation, its influencing factors, as well as the potential of place-responsive measures to nurture individual resilience. Obtained from quantitative and qualitative analyses, the assessment results indicate that Gongming has low individual resilience, failing to make up for its high risk of vulnerability to urban flooding.

The personal perceptions of individuals can explain the low resilience of residents of Gongming. These were embedded into their low perceived disaster awareness, poor perceived social environment among individuals, and perceived inefficient sewage systems, especially ecological systems. Based on the limiting factors, the problems of the present learning paths and loops, including individual learning, collective learning, and government learning, were diagnosed. Finally, the potential improvement of resilience performance through the current place-responsive measures associated with sponge city projects such as field visits to LID practices, Smart Sponge City for information sharing, NGO involvement, and the site selection for sponge bodies were explored and examined. The findings suggested that the high-level initiative of government couldn't realize resilience nurturing under the present state-centric model. Based on these results, place-responsive implications were identified and recommendations proposed. Finally, the contributions of this research to the literature on urban resilience were presented, followed by a discussion on the limitations of this research and recommendations for future studies.

[1] https://www.rand.org/health/surveys_tools/mos/36-item-short-form/scoring.html. http://www.measuringimpact.org/s4-sf-36.

Appendix A
Questionnaires

Questionnaire for residents

Respondents should be the residents of the study area (11 communities) and older than 16 years of age (included).

Personal and socioeconomic characteristics

1. Gender: 1☐Male 2☐Female
2. Age: 1☐16–18 (including18) 2☐18–40 (including 40)
   3☐40–60 (including 60) 4☐Older than 60
3. Married: 1☐Yes 2☐No
4. Education level: 1☐Up to middle school 2☐High school
   3☐Junior college or bachelor's degree 4☐Professional degree
5. *Hukou* status: 1☐Have Shenzhen *hukou* 2☐Do not have Shenzhen *hukou*
6. Occupation: 1☐Employed (Occupation: _______) 2☐Unemployed
   3☐At-home workers 4☐Full-time student
   5☐Part-time work and part-time school 6☐Retired 7☐Other_____
7. Working days per week: 1☐Less than 5 2☐5 3☐More than 5
8. Income (Unit: CNY/month): 1☐ $\leq$ 3000 2☐3000–7000 3☐7000–15000 4☐15000–30000 5☐ $\geq$ 30000
9. Name of living community: ______________________________
10. (a) Property ownership: 1☐Renter 2☐Owner 3☐Other ___________
    (b) Property type: 1☐Commercial property 2☐Apartment 3☐Small property housing (urban village) 4☐Others ______________________

J. Song, *Individual Resilience to Urban Flooding and the Implications for Urban Management*, https://doi.org/10.1007/978-981-97-0546-7

11. The primary means of commuting:
    1□Public transportation 2□Bike or walk 3□Private car 4□Other

Understand on urban flooding

1. Do you know urban flooding (waterlogging after heavy rain)?
   1□Yes 2□No
2. Have you ever suffered urban flooding?
   1□Yes 2□No
3. Have you lost anything in an urban flood?
   1□Yes 2□No

   If yes, what do you lost?
   1□Property 2□Production goods (facility, equipment and shop) 3□Vehicles
   4□Personal clothing 5□Other
4. Does the government offer you assistance and/or support after you lost something in an urban flood?
   1□Yes 2□No
5. Did some workplaces give you help and support after you lost something in an urban flood?
   1□Yes 2□No
6. How dangerous do you think urban flooding is?
   1□Very dangerous 2□Dangerous 3□Not very dangerous 4□Not dangerous at all
7. What are the dangers of urban flooding?
   1□Eeasy to fall down 2□Drowning or falling into a drain well
   3□Threat to children and elderly 4□Vehicle stalls in rain water
   5□Water flows into houses 6□Infrastructure (e.g., water, electricity, gas) breaks down 7□Other ______________
8. Does urban flooding affect your mental health?
   1□Yes 2□No
9. Do you know of other types of help for dealing with the urban flooding aside from the government and residents' committee?
   1□Yes 2□No

Personal capacity

1. Do you check the weather report every day?
   1□Yes 2□No
2. If you know that there will be a heavy rain tomorrow, will you tell your families or friends to prepare for it?
   1□Yes 2□No
3. How do you get weather information, especially for heavy rains? (3 is the path most commonly used, 2 is the path regular use, 1 is the path general use, 0 is the path have not been used)

Paths	None	General	Regular use	Most commonly used
WeChat	0	1	2	3
Text message	0	1	2	3
Websites	0	1	2	3
TV broadcast, newspaper	0	1	2	3
Community provision	0	1	2	3
Public lectures	0	1	2	3
Other: ________________				

4. Do you prepare for heavy rains?
   1□Yes 2□No
5. Have you ever learnt how to save yourself in an urban flood?
   1□Yes 2□No
6. Do you study common science on disaster and urban flooding prevention?
   1□Yes 2□No
7. Do you have the ability to help others escape when there is urban flooding?
   1□Yes 2□No
8. In a disaster, do you think you can help others or do you leave as soon as possible when there is urban flooding?
   1□Leave as soon as possible 2□Help others
9. Will you believe others when they want to help you?
   1□Yes 2□No
10. What paths would you like to use for learning about it?

Paths	None	General	Regular use	Most commonly used
WeChat	0	1	2	3
Text message	0	1	2	3
Websites	0	1	2	3
TV broadcast, newspaper	0	1	2	3
Community provision	0	1	2	3
Public lectures	0	1	2	3
Other: ________________				

11. Are you willing to help others when they are in need?
    1□Yes 2□No
12. Have you ever heard about sponge city?
    1□Yes 2□No
13. Can you tell us your understanding of sponge city?
    1□Yes (Please write down your understanding of it)________________
    ____
    2□No

14. Sponge city is an advanced concept on urban flooding prevention. Would you like to learn about it?
    1□Yes 2□No
15. Which paths do you want to use to get the knowledge?

Paths	None	General	Regular use	Most commonly used
WeChat	0	1	2	3
Text message	0	1	2	3
Websites	0	1	2	3
TV broadcast, newspaper	0	1	2	3
Community provision	0	1	2	3
Public lectures	0	1	2	3
Other: ______				

16. Have you ever received an invitation to participate in the management of urban flooding?
    1□Yes 2□No
17. Besides government, the improvement of capacity for fighting against urban flooding needs residents' participation. Are you interested?
    1□Yes 2□No
18. Besides government, the improvement of capacity for fighting against urban flooding needs residents' participation. Are you interested if you got paid from governments?
    1□Yes 2□No
19. Besides government, the improvement of capacity for fighting against urban flooding needs residents' participation. Are you interested if you got paid from governments?
    1□Yes 2□No 3□Unsure
20. Which paths do you want to choose for improvement the present situation for fighting against urban flooding in your living zone?
    1□Personal investment
    2□Participate into the planning and design for community against urban flooding 3□Participate into the community urban flooding management and popularization the concept to fight against urban flooding
    4□Participate into the construction projects of living community (e.g. green roof construction)
    5□Participate into the maintenance after the construction of infrastructures against urban flooding 6□Depends
21. Are you familiar with rainstorm warning signals?
    1□Don't know 2□Basically know 3□Know 4□Familiar
22. Do you follow the rainstorm warning signals strictly?
    1□Yes 2□No

Community/living zone prevention

1. According to your personal perceptions and answer the following questions:

Paths	None	General	Regular use	Most commonly used
WeChat	0	1	2	3
Text message	0	1	2	3
Websites	0	1	2	3
TV broadcast, newspaper	0	1	2	3
Community provision	0	1	2	3
Public lectures	0	1	2	3
Other: ________________				

2. Do you think your neighbors will help you when you are in danger?
   1□Yes 2□No
3. Does your community have some prevention strategies for urban flooding?
   1□Yes 2□No 3□Unsure
4. What do you think of the sense of well-being of your community?
   1□Very good 2□Good 3□General 4□Not good 5□Very good
5. Do you worry about your children when there is heavy rain?
   1□Yes 2□No 3□Unsure
6. Do you worry about your parents when there is heavy rain?
   1□Yes 2□No 3□Unsure
7. Are there many urban development projects around your community?
   1□Many 2□Average 3□Not many
8. In your mind, are there many ecological lands (grassland, forests, lakes) to be developed into buildings and roads around your community?
   1□Yes 2□No
9. Are you willing to have urban lands changed into the ecological lands?
   1□Yes 2□No 3□Other ________________
10. Do you think that your community's sewage system is efficient?
    1□Very efficient 2□Efficient 3□Average 4□Not very efficient 5□No efficient
11. What causes urban flooding?
    1□Blocked sewage pipes 2□Old sewage systems
12. What do you think will block the sewage system?
    1□Dry branches and fallen leaves 2□Mud 3□Gravel 4□Household garbage 5□Other ________
13. How clean is the living environment of your community?
    1□Very clean 2□Clean 3□Average 4□Not very clean 5□Not clean
14. Does your community dispose of its garbage efficiently?
    1□Very efficient 2□Efficient 3□Average 4□Not very efficient 5□No efficient

15. Do you know if somebody will clean the blocked pipes? Who?
    1☐Yes 2☐No 3☐Unsure
    1☐Residents 2☐Sanitation worker 3☐City management 4☐Other ___
    ______
16. How long does it take to clear the blocked pipes?
    1☐Less than 1 h 2☐6 h 3☐12 h 4☐24 h 5☐More than 24 h
17. Do you know if somebody maintains the sewage system in your living zone?
    1☐Yes (______________________) 2☐No 3☐Unsure
18. How often do they maintain sewage system in your living zone?
    1☐One week 2☐One month 3☐One season 4☐Half a year 5☐One year 6☐More than one year 7☐Never 8☐Do not know
19. Does your community have other types of drainage strategies?
    1☐Yes 2☐No 3☐Unsure

Appendix B
Interview Questions and Interviewee List

Questions for residents

1. Does your community have lectures on urban flooding prevention or disaster prevention?
2. Do you think urban flooding is dangerous?
3. Do you know where have serious urban floods around your community or your workplace? Do shop or factory owners lose anything in the floods?
4. Do you trust yourself, people around you, your community and government? Do you want to invest in mitigating the damage of urban flooding?
5. Does your residents' committee do anything to reduce the threat of urban floods?
6. Do you feel the situation of urban flooding has improved? Do you know the reasons?
7. Have you ever seen the government come to your community to improve the present situation about urban floods?
8. Do you know the management of Maozhou River?
9. Have you ever heard of Sponge City?

Questions for officials

1. To what degree does Shenzhen cooperate with different departments on urban flooding management?
2. Dose Shenzhen have special planning for urban flooding management?
3. How does the government cooperate with industries and residents for urban flood management?
4. Do you think it is necessary for public participation on urban flooding management?
5. Do you have any management focusing on the improvement of social resilience capacity of individuals against urban floods?
6. What is the sponge city construction going?
7. How does government balance urban development and the decrease of natural water system?

J. Song, *Individual Resilience to Urban Flooding and the Implications for Urban Management*, https://doi.org/10.1007/978-981-97-0546-7

8. Do you have some websites to check the urban flooding situations?
9. Do you have some lectures to deliver to residents for disaster prevention and personal safety?
10. Do you have some paths for checking the urban flood points around our living zones?

Questions for developers and architect

1. What do you think of green buildings?
2. Do you think it is necessary to build green buildings?
3. How does urban planning balance the profits and green ratio during development?
4. Are developers willing to add LID concepts into the development?
5. Do you know about sponge city? Do you know how to add sponge city to the development?

Questions for planners

1. How is sponge city construction going, especially in its planning and design?
2. What are major problems with sponge city construction?
3. Do you think it will have the predicted effects?
4. How do you choose where to build the sponge bodies?
5. Why did you choose Guangming New District as a pilot point?
6. Do you have public participation in the planning process?
7. Do you focus on the residents when making plans and designs?
8. Do you factor the resilience concept into your planning?

Questions for residents' committee and managers of residential zone

1. Do you have strategies to help the residents escape from heavy rain and urban floods?
2. Are there some shortcomings with your management of urban floods?
3. Is there a sense of community where you live? Do the residents cooperate with your work?
4. What are the major difficulties in terms of promoting coping and adaptive capacities for urban floods?

Questions for NGOs

1. What have you done to help residents cope with and adapt to urban flooding events?
2. What are you willing to do for residents to help them cope with and adapt to urban flooding events?
3. Do you mainly serve Government or Residents?
4. What are your funding sources to carry out your plans?
5. What are the limitations for you to carry out what you want to do, e.g. help residents to cope with and adapt to urban flooding?

Questions for scholars

1. What do you think of the strategy of sponge city construction in China and Shenzhen?
2. What do you think of water management in Shenzhen?
3. Do you have some suggestions to improve the urban flooding management from the perspective of disciplinary development?

Interviewee list

Type	Number	Basic characters
Gongming	1	Male, young
Gongming	2	Male, elderly
Gongming	3	Male, elderly
Jiangshi	4	Male, young
Jiangshi	5	Female, elderly
Shangcun	6	Male, young
Shangcun	7	Female, young
Xiacun	8	Male, elderly
Xiacun	9	Female, young
Heshuikou	10	Male, young
Heshuikou	11	Male, elderly
Citianpu	12	Male, young
Genzhuyuan	13	Male, young
Yulv	14	Male, young
Yulv	15	Female, young
Tianliao	16	Female, young
Officials	A	Working in the municipal government (planning)
	B	Working in the municipal government (meteorology)
	C	Working in the Guangming New District government
	D	Working in the Guangming New District government
Developers	A	Working in CapitaLand China
	B	Working in Overseas Chinese Town (OCT)
	C	Working in China Overseas Land and Investment LTD.
	D	Working in China Merchants Land Limited
Planners	A	Working in the planning institute whose specialty is municipal engineering
	B	Working in the planning institute whose specialty is municipal engineering
	C	Working in the planning institute whose specialty is landscape ecology

(continued)

(continued)

Type	Number	Basic characters
	D	Working in the planning institute whose specialty is urban flooding management
	E	Working in building research institute whose specialty is green building
Managers of residential zone	A	Working as a housing manager of a commercial property
	B	Working as a housing manager of a urban village
Scholars	A	Research on water science
	B	Research on governance, management and urban planning
Residents' committee	A	Working in the residents' committee of Yulv
	B	Working in the residents' committee of Gongming
NGOs	A	NGO member focusing on water environment protection
	B	NGO member focusing on sponge city construction
	C	NGO member focusing on natural pollution control

Appendix C
Drainage Capacity Evaluation

Drainage capacity in this research is evaluated based on the model of SWMM and the steps for building SWMM are described in the following sections.

(1) **Sewerage system generalization of study area**

The process of water accumulation and drainage is complicated in urban areas. The hydrological properties are quite different for permeable and impermeable surfaces. Even for permeable surfaces, different land use types show different properties. The diverse and intricate surfaces with uneven properties in urban areas make it difficult to perform realistic simulations. The sewerage systems of big cities have complicated and fragmented distribution systems, which further increase the difficulty to simulate the accumulation and drainage processes of rainwater. Therefore, it is necessary to simplify and generalize the land use and the sewerage system of the study area.

Underlying surface generalization

The study area covers 37.68 km^2. Using the land use map and the hydrographic features, the study area is divided into grass, forest, farmland, bare lands, gardens and green lands, vacant lands, water areas, low-density construction lands (industrial lands and public facilities lands), high-density construction lands (urban construction lands and wholesale and retail lands), and transportation lands. The generalized land use map is shown as Fig. C.1.

Pipes and pipe junctions generalization

The number of sewerage pipes and pipe junctions is too large for the model therefore they should be simplified for the simulation. The rules for pipe and junction generalization are as follows:

a. The minimum distance between the two junctions is 100 m,
b. Keep the corner junctions, changing diameter junctions, or large variation range of slope junctions,

J. Song, *Individual Resilience to Urban Flooding and the Implications for Urban Management*, https://doi.org/10.1007/978-981-97-0546-7

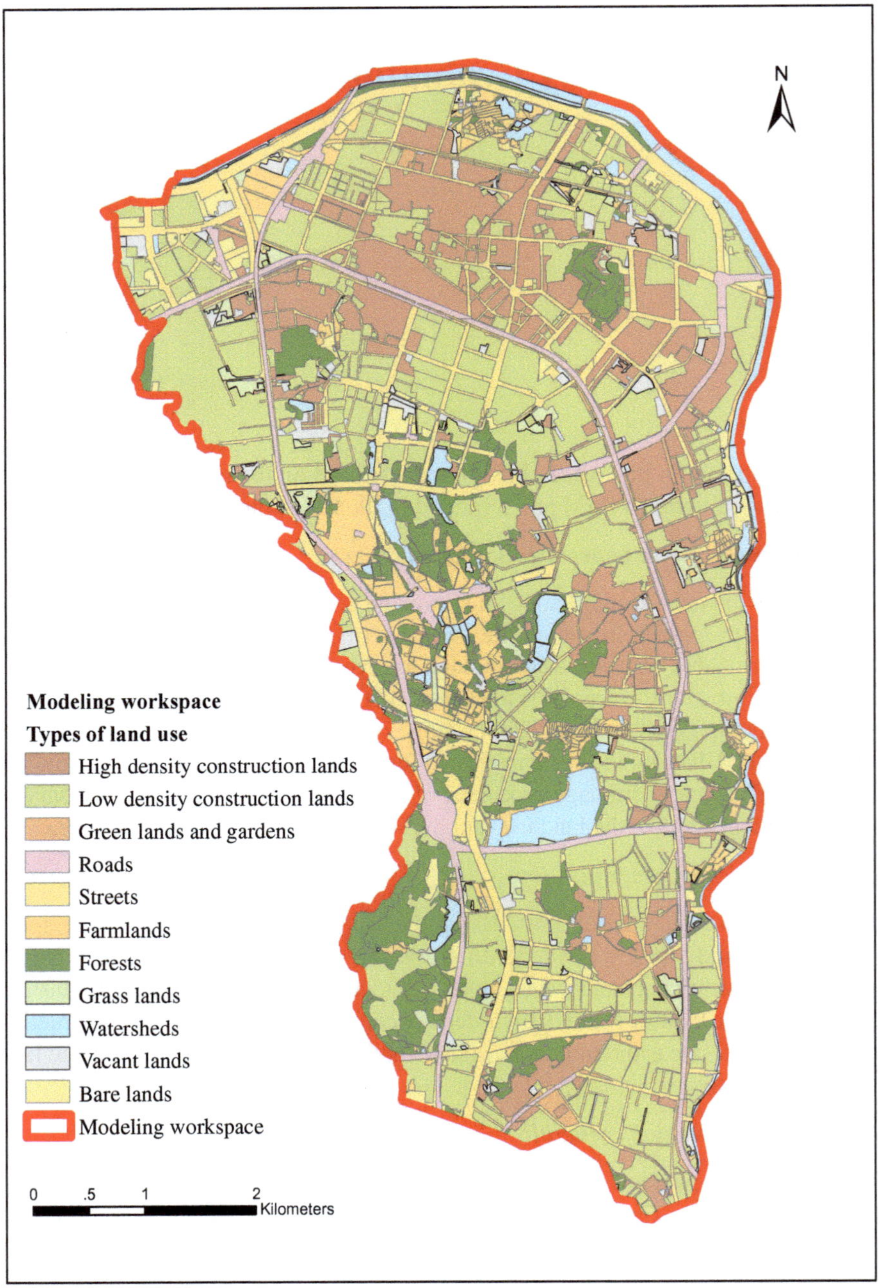

Fig. C.1 Generalization map of land use. *Source* Author

c. Keep the parallel pipes and junctions on both sides of the roads,
d. Delete the spur pipelines of roads and try to keep the drainage capacity of pipelines the same during the generalization.

Original distributions of pipelines and junctions have two sources; one is from the land use map of Shenzhen (shp) and the other is from sewerage system project map (dwg). Based on the rules and the original base maps, all the pipelines and junctions of study area were simplified to 1151 junctions, including 56 outlets, 1095 inspection wells, and 1044 pipelines, including 49 gouges. Based on the slope and depth of the pipeline junctions, we calculated the flow directions of different pipelines, and the coordinates and length of the pipelines using the module of Calculate Geometry of ArcGIS10.2. Finally, the spatial distribution of pipelines and pipe junctions was simplified as shown in Figs. C.2 and C.3.

Catchment partition

DEM (Fig. C.4) is the original data to determine the catchment partition. Using the module of Spatial Analyst Tools (Surface) of ArcGIS 10.2, DEM was completed, slope and flow direction were calculated, and the flow path of water, river networks, and the basins of the study area were obtained. Finally, the basin partition was calculated.

The extracted basins cannot exactly reflect the properties of the underlying surfaces of the study area because the precision of DEM is limited. Since most of the study area includes developed land rather than natural lands, the basin partition process based only on the DEM is not very accurate. Therefore, it is necessary to perform further manual repartition by adding the file of the statutory plans of the study area to ArcGIS 10.2 and combining the partition results of basins with the repartition of the catchments of the study area. Finally, considering the land use types, zoning distributions, and outfalls of the junction distributions, the study area was divided into 1116 catchments (Fig. C.5).

The catchments, pipelines, pipe junctions were added into SWMM with the spatial coordinates and the study area in SWMM are shown in Fig. C.6.

(1) **Parameter setting and calculation**

There are two types of parameters in SWMM (Table C.1): measurement parameters and calibration parameters. Measurement parameters can be calculated by ArcGIS, while the calibration parameters are the ones within a range of reasonable interval and determined through model calibration.

Calculation of measurement parameters

Area, length, coordinates X and Y can be calculated using the Calculate Geometry Tool in ArcGIS. Average slope (%) (Fig. C.7) of the study area can be calculated using the module of Spatial Analyst Tools (Zonal Statistics) of ArcGIS based on DEM.

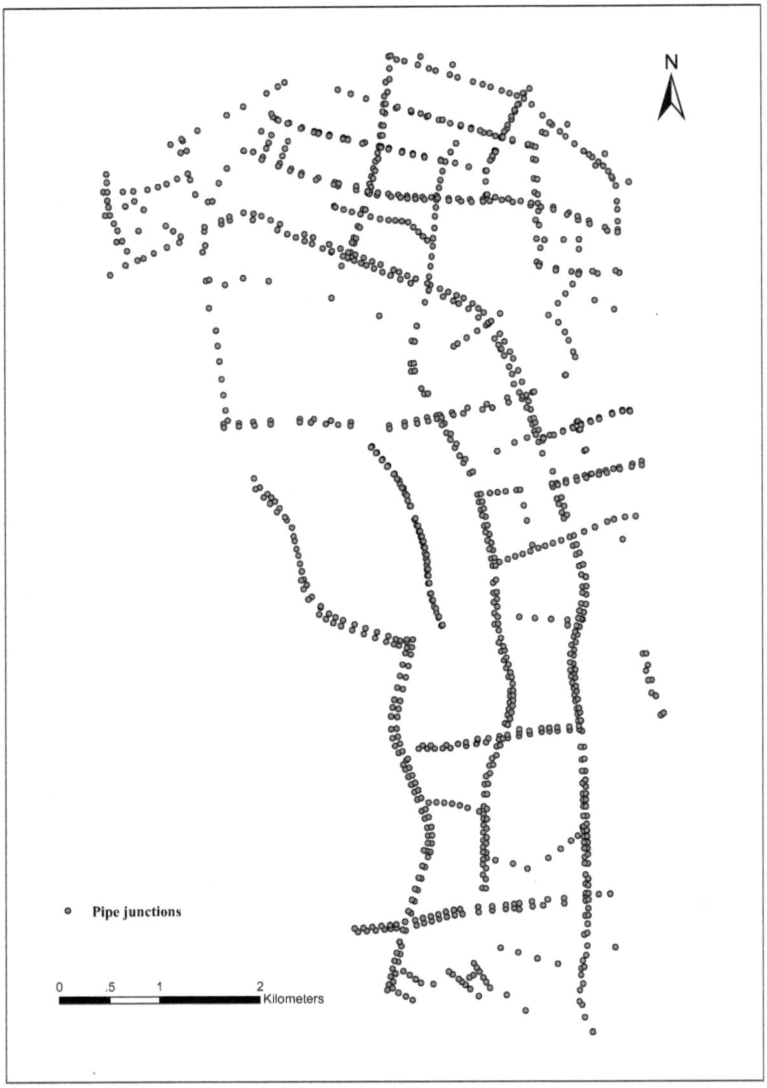

Fig. C.2 Generalization of pipe junctions. *Source* Author

Impermeability depends on the properties of the underlying surfaces of the study area. In this study, the land use map of 2014 is selected and simplified. As shown in Fig. C.1, there were 11 types of land use was generalized for our research area. The impermeability coefficient is valued based on different land use types (Table C.2). The Analyst Tools module (Summary Statistics) of ArcGIS was selected to calculate the impermeability coefficient of different catchments based on the weighted average area of each type of land use.

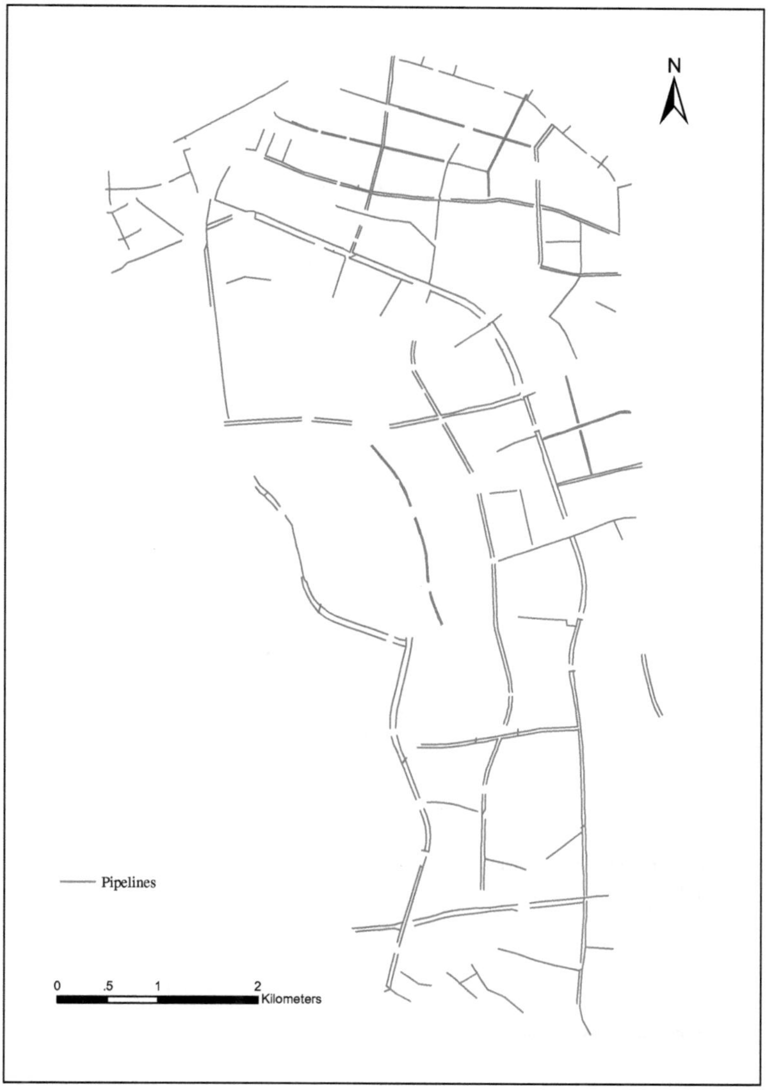

Fig. C.3 Generalization of pipelines. *Source* Author

Calculation of the calibration parameters

According to the user manual of the SWMM model,[1] the ranges of the different calibration parameters are shown in Tables C.3 and C.4.

Considering the true conditions of the study area, such as climate, soil type, underlying surfaces, and pipeline materials, the parameters in this analysis are limited to the ranges in Table C.5.

[1] https://nepis.epa.gov/Exe/ZyPDF.cgi?Dockey=P100NYRA.txt.

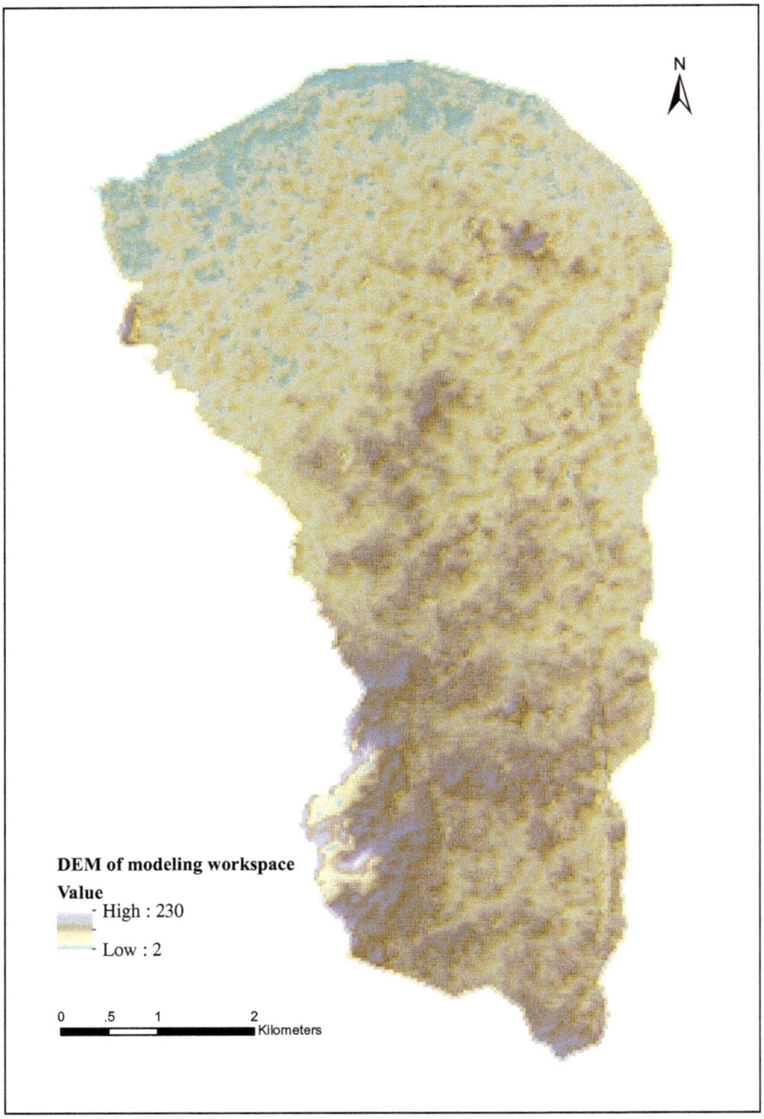

Fig. C.4 DEM of modeling workspace. *Source* Author

Among these parameters, "width" plays a dominant role on the accuracy of the model simulation. Width is defined as the ratio between the catchment area and the longest length of the overland flow. In practice, the process of water accumulation and drainage is complicated and the paths of overland flow are difficult to calculate. Therefore, the width cannot be truly measured. The four methods for the indirect measurement of width are listed as follows (Sun, 2011).

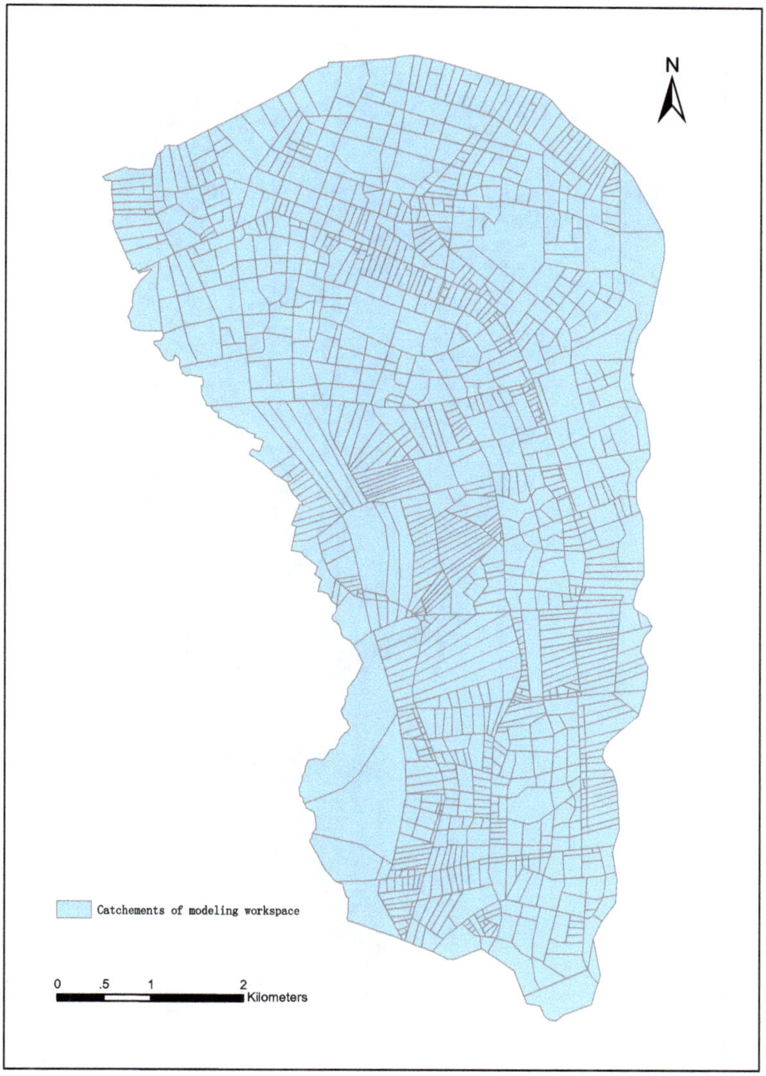

Fig. C.5 Catchments of modeling workspace. *Source* Author

a. Width = 1.7*MAX (Height, Width);
b. Width = K*Sqrt (Area) (0.2 < K < 5);
c. Width = K*Perimeter (0 < K < 1);
d. Width = Area/Flow Length.

Among these four methods, b and d are quite commonly used. In this study, all of the four methods have been tested and finally, method b was chosen to calculate the width because the simulation results of b were more accurate.

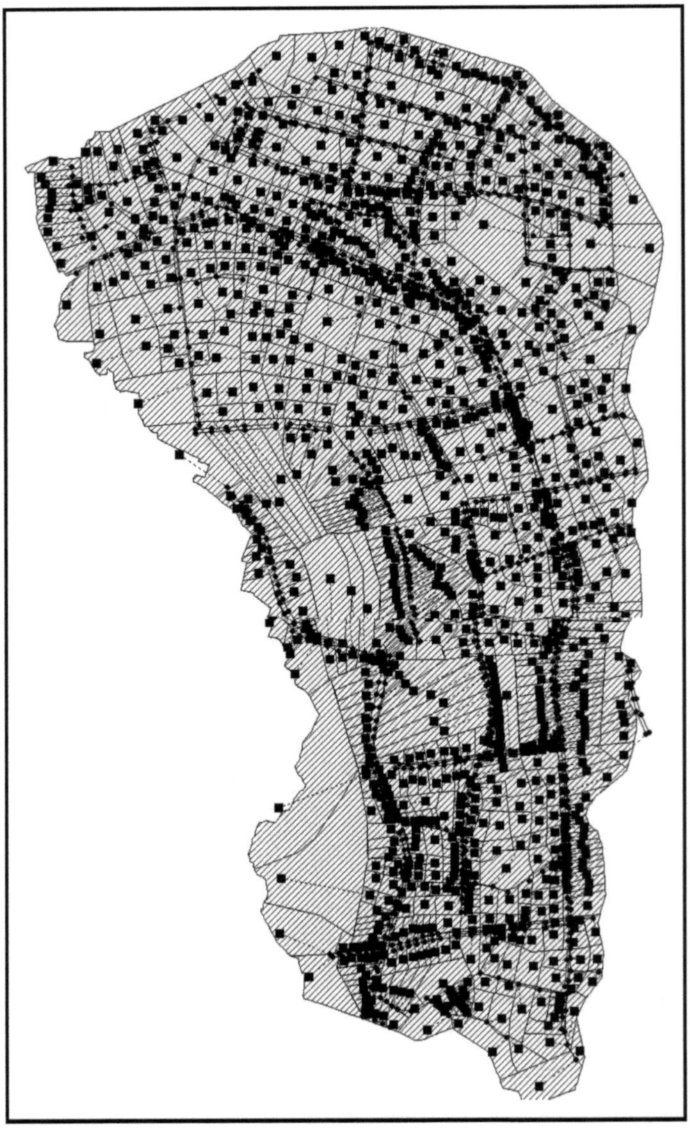

Fig. C.6 Workspace of research area in SWMM. *Source* Author

After building the SWMM model and testing its accuracy, the duration of each urban flooding area can be directly read through the result reports of SWMM.

(2) **Drainage capacity performance**

Duration is the length of time from the apperance of waterlogging to the fading of waterlogging. It can be determined using the simulation results of SWMM. The precipitation process on 11 May 2014 was used to simulate the urban flooding

Table C.1 Parameter types of SWMM

Parameter types	Measurement parameters	Calibration parameters
Catchments	Area, slope, impermeability	Width, roughness coefficient of impermeability, roughness coefficient of permeability, surface-depression storage of impermeability, surface-depression storage of permeability
		Zero surface-depression storage of impermeability
Conduits	Shape, maximum depth, length	Roughness coefficient
Junctions	The inner bottom elevation, maximum depth, flooded area	None
HORTON infiltration models	None	Maximum infiltration rate, Minimum infiltration rate, decay constant, drying time

Source Reference to the website of SWMM (https://www.epa.gov/water-research/storm-water-management-model-swmm) and research of Knighton et al. (2016)

process, and then the measured data on depth and spatial distribution of urban flooding (on the same day) were used to calibrate the parameters and validate the simulation. The spatial distribution of test points and their values are shown in Fig. C.8 and Table C.6. The simulation results of the depth and spatial distribution of urban floods are shown in Fig. C.9. There were a total of 12 points, and six of them were applied to calibrate the parameters of SWMM; the other six were used to validate the simulation results (Table C.6). The calibration and validation data are shown in Table C.7. The existing literature shows that in foreign countries, such as in the city of Cincinnati (Ohio, USA), ± 15% was used as the tolerance level of model accuracy for peak depth (Siegrist et al., 2016); while in China, some scholars pointed out that 20% would be acceptable, given the relatively large scale of the simulations (Yu et al. 2012). Considering that this research area is quite large, and that the terrain is complicated, the tolerance level was set equal to or less than 30%. The results demonstrate that the SWMM meets Gongming's most recent modeling standards. The results basically fit the requirements of simulation, but with relative error spread from 1 to 30%. Some of the calibration parameters are shown in Table C.8.

The final model was built based on the calibration parameters, and based on the model using the rainfall data of 11 May 2014, the duration could be read from the model. The spatial distribution of the simulated duration results is shown in Fig. C.10. It should be mentioned that the parameter 'duration' is different from that called 'flooding hours' shown in Table C.9. The parameter 'Flooding hours', shows the number of hours at peak depth, while 'duration' is the length of time from the apperance of waterlogging to the fading of waterlogging, which can be read directly from the simulation. Based on the simulated duration results, we were able to find that for most of the places that appeared waterlogged on that day in the resarch area; the

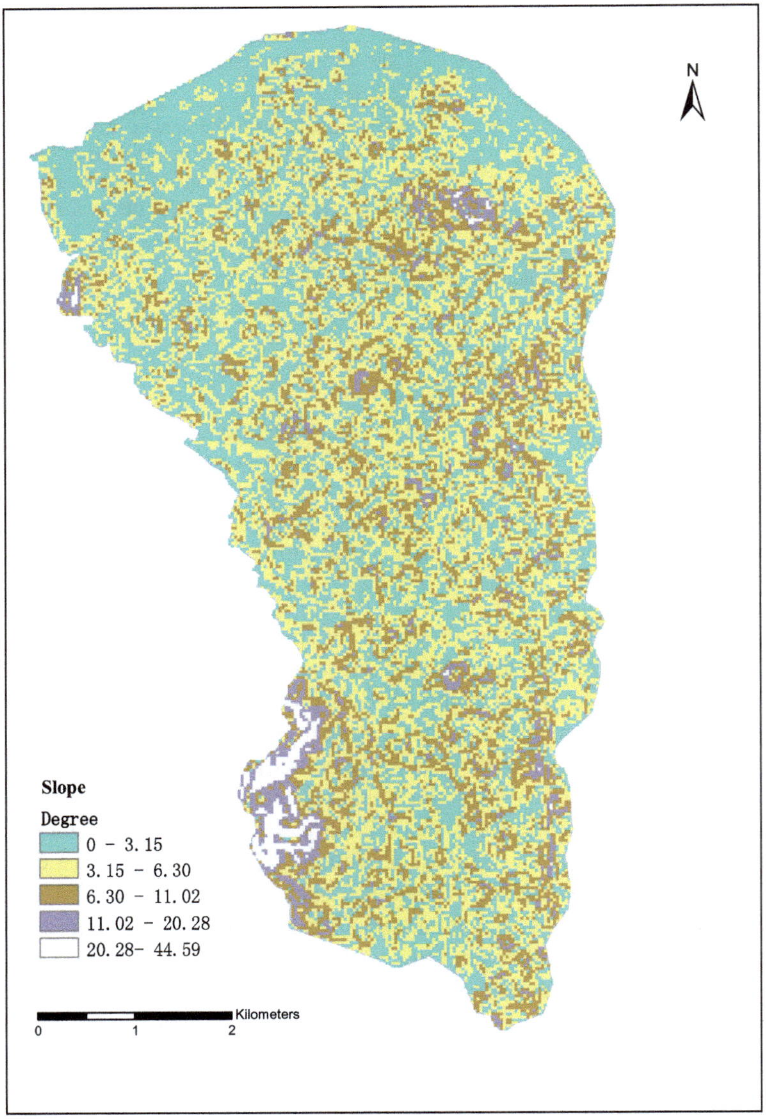

Fig. C.7 Slope (%) of the research area. *Source* Author

duration was from 0 to 21.83 h. We found that there were mainly two long duration belts in this area: one in the northern part of the research area, and the other in the middle of the research area, in a densely populated urban area.

Table C.2 Impermeability coefficients of different land use types

Types	Permeability			
Name	Grass, forest, farmland, bare lands, garden and green lands, vacant lands, water areas	Low-density construction lands	High-density construction lands	Transportation lands
Impermeability coefficient	0	0.8	0.9	1

Source Cite from Sun (2011)

Table C.3 Depression storages of different types of underlying surfaces

Underlying surfaces	Depression storage/mm
Impermeable surface	1.27–2.54
Grass	2.54–5.08
Pasture	5.08
Forest	7.62

Source Cite from Qin et al. (2013) and Sun (2011)

Table C.4 Range of manning coefficient (N value)

Underlying surfaces	N	Pipe types	N
Smooth-surfaced asphalt	0.011	Asbestos-cement pipe	0.011–0.015
Smooth-surfaced concrete	0.012	Bricked pipe	0.013–0.017
Conventional concrete	0.013	Casting ductile iron pipe	0.011–0.015
Wood	0.014	Concrete (entirely)	
Clay	0.015	Smooth-surfaced	0.012–0.014
Fallow soil (no vegetal residues)	0.050	Unsmooth-surfaced	0.015–0.017
Cultivated soil (residual vegetation < 20%)	0.060	Corrugated metal pipe	
Cultivated soil (residual vegetation > 20%)	0.170	Flat surface	0.022–0.026
Pasture (nature)	0.130	Paved inner sole	0.018–0.022
Grass	0.150–0.410	Fiber-reinforced asphalt lining	0.011–0.015
Forest	0.40–0.80	Plastic conduit (smooth-surfaced)	0.011–0.015
		Concrete pipes	0.011– 0.015

Source Cite from Qin et al. (2013) and Sun (2011)

Table C.5 Range of calibration parameter of the study area

Parameter	Value range
Width/m	K*SQRT (area)
N-impervious	0.010–0.015
N-pervious	0.10–0.30
Depression storage-impervious/mm	2–5
Depression storage-pervious/mm	3–10
Zero-depression storage-impervious/%	5–30
Roughness of pipeline	0.013–0.015
Maximum infiltration rate/(mm/h)	25.4–254
Minimum infiltration rate/(mm/h)	0.3–120.4
Decay constant	2–7
Drying time	2–14

Source Reference to some scholars' research, such as Sun (2011) and Qin et al. (2013)

Duration of urban flooding can reflect the drainage capacity of urban drainage systems, with longer duration indicating low drainage capacity. To dichotomize duration, it was equally divided into four continuous arrays based on original data, and a score was assigned to each array, and then drainage capacity could be scored (Table C.9). The drainage capacity was spatially linked with the living places of the residents that answered the questionnaire. The spatial distribution of the domiciles of these residents was identified using the names of their residence zones. Then, making use of the spatial analyst tool ArcGIS, the drainage capacities were extracted to the geometric central points of the residence zones. Finally, the drainage capacity in the area nearest each resident (according to their residence zones) was generated.

Table C.6 Measurements of urban flooding points for model testing

ID	County	Location	Longitude	Latitude	Area	Depth	Flooding hours
1	Shangcun	Road crossing of Minsheng Street and Xingfa Road	113.903	22.785	1000	0.25	4
2	Shangcun	Road crossing of Nan Ringroad and Bieshu Road	113.907	22.780	2000	0.30	4
3	Shangcun	Road crossing of North Ringroad and Changchun North Road	113.895	22.798	2000	0.30	3
4	Gongming	Road crossing of Huafa North Road and Minsheng Dadao	113.903	22.786	5000	0.20	3
5	Jiangshi	Outside of Yihemoju on Genyu Road	113.887	22.731	3000	0.60	4
6	Jiangshi	Jiangshi Road section of Genyu Road	113.885	22.748	2000	0.50	1
7	Tangwei	Zhoujia Dadao on Songbai Road	113.899	22.764	3000	0.25	3
8	Xiacun	Road crossing of West Ringroad and Keyu Road	113.865	22.791	1000	0.20	4
9	Tianliao	Outside of Weidong Store	113.900	22.729	2000	0.30	3
10	Mashantou	Auto market of Gongming	113.870	22.787	500	0.20	2
11	Jiazitang	Gate of Jiazitang County	113.908	22.740	500	0.25	2

Source Guangming Wind, Flood, and Drought Prevention Office (2016)

Table C.7 Calibration and validation results

Points ID	Observed (m)	Modeled (m)	Errors (m)	% Difference
Calibration				
1	0.25	0.255	0.005	2
2	0.30	0.285	– 0.015	– 5
3	0.30	0.320	0.020	6
4	0.20	0.217	0.017	8
5	0.20	0.224	0.024	11
6	0.50	0.453	– 0.047	– 10
Validation				
7	0.50	0.441	– 0.059	– 13
8	0.25	0.248	– 0.002	– 1
9	0.20	0.174	– 0.046	– 30
10	0.30	0.205	– 0.085	– 41
11	0.20	0.165	– 0.035	– 18
12	0.25	0.226	– 0.024	– 11

Source Author

Table C.8 Values of some calibration parameters

Parameters	Calibration value
K	2
N-impervious	0.015
N-pervious	0.15
Depression storage-impervious/mm	5
Depression storage-pervious/mm	15
Zero-depression storage-impervious/%	25
Roughness	0.013
Maximum infiltration rate/(mm/h)	76
Minimum infiltration rate/(mm/h)	12
Decay constant	2
Drying time	5

Source Author

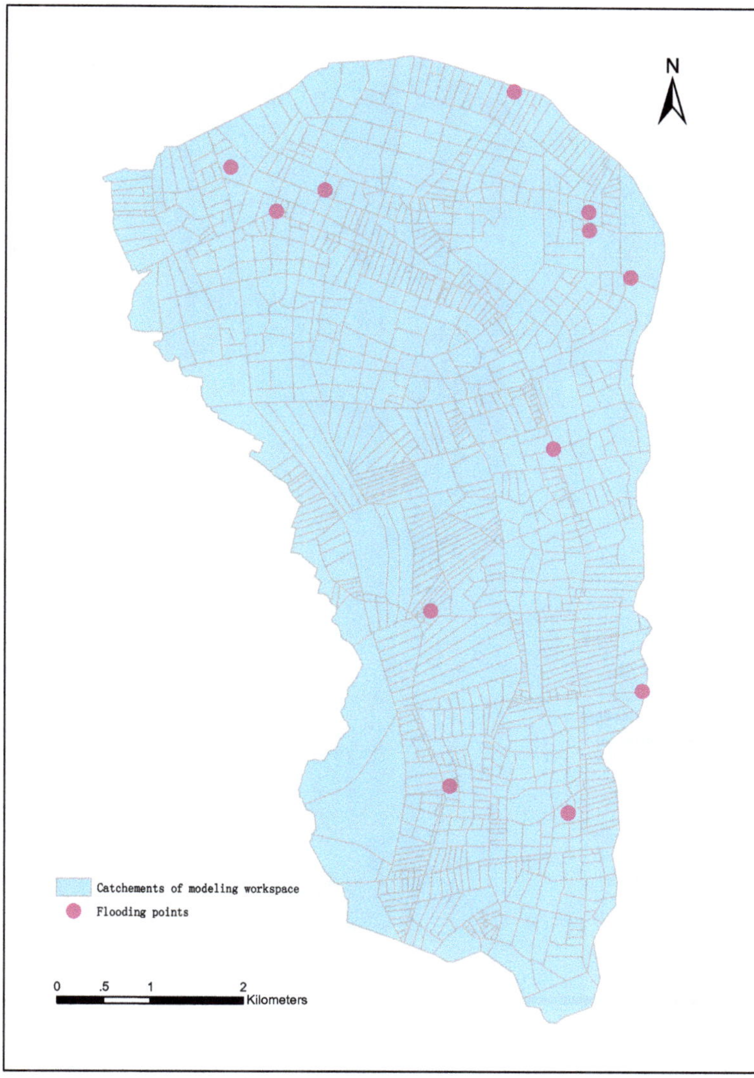

Fig. C.8 Spatial distribution of the test points of urban flooding in the research area. *Source* Author

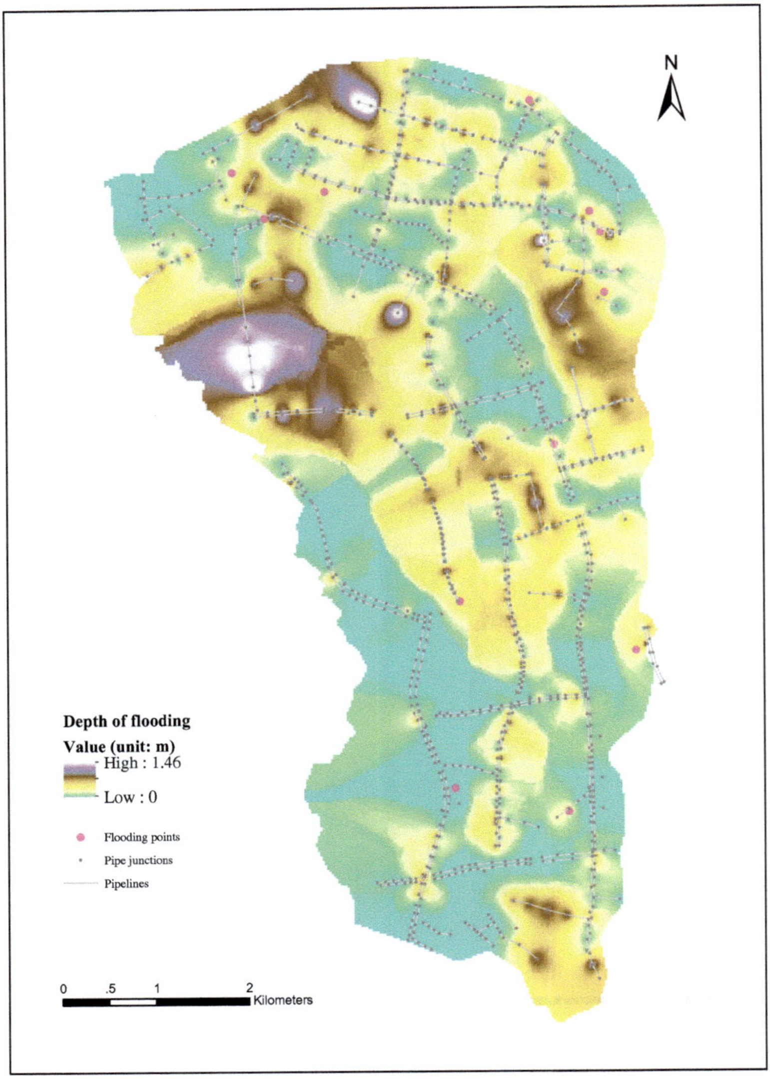

Fig. C.9 Spatial distribution of depth of urban flooding. *Source* Author

Table C.9 Scoring for drainage capacity

Duration (hours)	0–4.33	4.33–8.66	8.66–13.00	13.00–17.33	17.33–21.66
Score of drainage capacity	10	7.5	5	2.5	0

Source Author

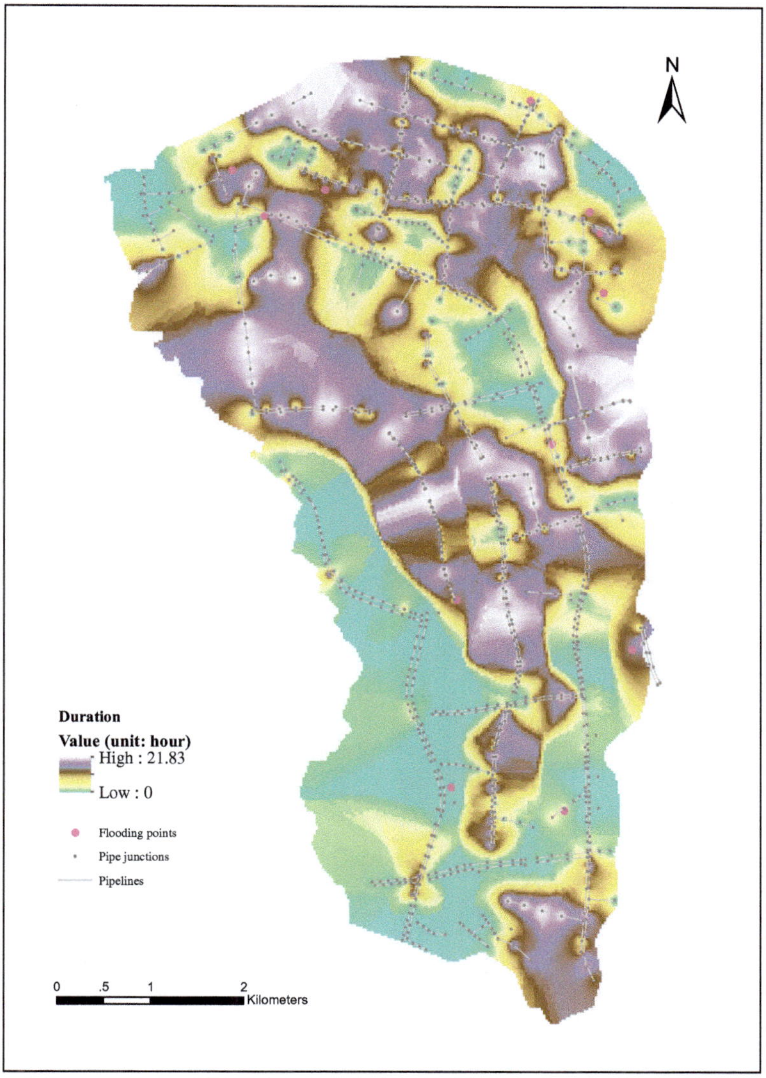

Fig. C.10 Spatial distribution of drainage capacity. *Source* Author

Appendix D

J. Song, *Individual Resilience to Urban Flooding and the Implications for Urban Management*, https://doi.org/10.1007/978-981-97-0546-7

Pictures

References

Abramson, D. M., Grattan, L. M., Mayer, B., Colten, C. E., Arosemena, F. A., Bedimo-Rung, A. and Lichtveld, M. (2015). The resilience activation framework: a conceptual model of how access to social resources promotes adaptation and rapid recovery in post-disaster settings. *The journal of behavioral health services & research, 42*(1), 42–57.

Adger, W. N. (2003). Social capital, collective action, and adaptation to climate change. *Economic geography, 79*(4), 387–404.

Adger, W. N., Hughes, T. P., Folke, C., Carpenter, S. R., and Rockström, J. (2005). Social-ecological resilience to coastal disasters. *Science, 309*(5737), 1036–1039.

Adger, W. N. and Vincent, K. (2005). Uncertainty in adaptive capacity. *Comptes Rendus Geoscience, 337*(4), 399–410.

Ahern, J. (2011). From fail-safe to safe-to-fail: Sustainability and resilience in the new urban world. *Landscape and urban planning, 100*(4), 341–343.

Ahern, J. (2013). Urban landscape sustainability and resilience: the promise and challenges of integrating ecology with urban planning and design. *Landscape Ecology, 28*(6), 1203–1212.

Ahern, M., Kovats, R. S., Wilkinson, P., Few, R. and Matthies, F. (2005). Global health impacts of floods: epidemiologic evidence. *Epidemiologic reviews, 27*(1), 36–46.

Ahrens, J. and Rudolph, P. M. (2006). The importance of governance in risk reduction and disaster management. *Journal of contingencies and crisis management, 14*(4), 207–220.

Alfieri, L., Burek, P., Dutra, E., Krzeminski, B., Muraro, D., Thielen, J. and Pappenberger, F. (2013). GloFAS-global ensemble streamflow forecasting and flood early warning. *Hydrology and Earth System Sciences, 17*(3), 1161.

Allen, C. R., Angeler, D. G., Garmestani, A. S., Gunderson, L. H. and Holling, C. S. (2014). Panarchy: theory and application. *Ecosystems, 17*(4), 578–589.

Allen, P. M. (2012). *Cities and regions as self-organizing systems: models of complexity*: Routledge.

Alshehri, S. A., Rezgui, Y. and Li, H. (2015). Delphi-based consensus study into a framework of community resilience to disaster. *Natural Hazards, 75*(3), 2221–2245.

Altman, I. and Werner, C. M. (2013). *Home environments* (Vol. 8): Springer Science & Business Media.

Amos, F. (1989). Strengthening municipal government. *Cities, 6*(3), 202–208.

Amri, A., Bird, D., Ronan, K., Haynes, K. and Towers, B. (2016). Disaster risk reduction education in Indonesia: challenges and recommendations for scaling up. *Natural Hazards and Earth System Sciences, 17*(4), 595–612.

Andjelkovic, I. (2001). *Guidelines on non-structural measures in urban flood management*. International Hydrological Programme (IHP), United Nations Educational, Scientific and Cultural Organization (UNESCO).

J. Song, *Individual Resilience to Urban Flooding and the Implications for Urban Management*, https://doi.org/10.1007/978-981-97-0546-7

Armitage, D. (2006). *Resilience management or resilient management? A political ecology of adaptive, multi-level governance.* Paper presented at the 11th Biennial Conference of IASCP held in Bali.

Arnbjerg-Nielsen, K., Willems, P., Olsson, J., Beecham, S., Pathirana, A., Gregersen, I. B., ... and Nguyen, V. T. V. (2013). Impacts of climate change on rainfall extremes and urban drainage systems: a review. *Water Science and Technology, 68*(1), 16–28.

Arnell, N. W. and Gosling, S. N. (2016). The impacts of climate change on river flood risk at the global scale. *Climatic Change, 134*(3), 387–401.

Arouri, M., Nguyen, C. and Youssef, A. B. (2015). Natural disasters, household welfare, and resilience: evidence from rural Vietnam. *World development, 70*, 59–77.

Asif, M., Sharma, R. B. and Adow, A. H. E. (2015). An empirical investigation of the relationship between economic growth, urbanization, energy consumption, and CO_2 emission in GCC countries: a panel data analysis. *Asian Social Science, 11*(21), 270.

Aulia, D. N. and Ismail, A. M. (2016). The Criteria of Residential Satisfaction in Gated Community: Medan City. *Asian Journal of Behavioural Studies, 1*(3), 43–52.

Aven, T. (2011). On some recent definitions and analysis frameworks for risk, vulnerability, and resilience. *Risk Analysis, 31*(4), 515–522.

Bajek, R., Matsuda, Y. and Okada, N. (2008). Japan's Jishu-bosai-soshiki community activities: analysis of its role in participatory community disaster risk management. *Natural Hazards, 44*(2), 281–292.

Baker, V. R. (1977). Stream-channel response to floods, with examples from central Texas. *Geological Society of America Bulletin, 88*(8), 1057–1071.

Beatley, T. and Newman, P. (2013). Biophilic cities are sustainable, resilient cities. *Sustainability, 5*(8), 3328–3345.

Belle, D., Smith-Doerr, L. and O'Brien, L. M. (2014). Gendered networks: professional connections of science and engineering faculty. In *Gender Transformation in the Academy.* Emerald Group Publishing Limited, 153–175.

Béné, C., Frankenberger, T. and Nelson, S. (2015). *Design, monitoring and evaluation of resilience interventions: conceptual and empirical considerations.* Working paper of IDS (Institute of Development Studies) on volume 2015, No.459.

Bennett, K. (1996). Measuring health state preferences and utilities: rating scale, time trade-off and standard gamble techniques. *Quality of Life Pharmacoeconomics in Clinical Trials*, 253–265.

Benson, C., Twigg, J. and Rossetto, T. (2007). *Tools for mainstreaming disaster risk reduction: guidance notes for development organisations*: Provention Consortium.

Berkes, F. (2007). Understanding uncertainty and reducing vulnerability: lessons from resilience thinking. *Natural Hazards, 41*(2), 283–295.

Berkes, F., Colding, J. and Folke, C. (2002). *Navigating social-ecological systems: building resilience for complexity and change*: Cambridge University Press.

Berkes, F., Folke, C. and Colding, J. (2000). *Linking social and ecological systems: management practices and social mechanisms for building resilience*: Cambridge University Press.

Berkes, F. and Turner, N. J. (2006). Knowledge, learning and the evolution of conservation practice for social-ecological system resilience. *Human Ecology, 34*(4), 479–494.

Berkes, F. and Ross, H. (2013). Community resilience: toward an integrated approach. *Society & Natural Resources, 26*(1), 5–20.

Bhatt, G., Sinha, K., Deka, P. and Kumar, A. (2014). Flood hazard and risk assessment in Chamoli district, Uttarakhand using satellite remote sensing and GIS techniques. *International Journal of Innovative Research in Science, Engineering and Technology, 3*(8), 15348–15356.

Bino, G., Kingsford, R. T. and Jenkins, K. (2014). Climate adaptation and adaptive management planning for the Macquarie Marshes: a wetland of international importance. *Applied Studies in Climate Adaptation*, 95–106.

Bohensky, E. L. and Leitch, A. M. (2014). Framing the flood: a media analysis of themes of resilience in the 2011 Brisbane flood. *Regional Environmental Change, 14*(2), 475–488.

Booth, D. B. (1991). Urbanization and the Natural Drainage System-Impacts, Solutions, and Prognoses. *The Institute for Environmental Studies, 7*(1), 93–118.

Brenner, N. (1999). Globalisation as reterritorialisation: the re-scaling of urban governance in the European Union. *Urban Studies, 36*(3), 431–451.

Brown, K. and Westaway, E. (2011). Agency, capacity, and resilience to environmental change: lessons from human development, well-being, and disasters. *Annual review of environment and resources, 36*, 321–342.

Bruneau, M., Chang, S. E., Eguchi, R. T., Lee, G. C., O'Rourke, T. D., Reinhorn, A. M., ... and Von Winterfeldt, D. (2003). A framework to quantitatively assess and enhance the seismic resilience of communities. *Earthquake spectra, 19*(4), 733–752.

Buckle, P. (2006). Assessing social resilience. In *Disaster resilience: An integrated approach.* Charles C Thomas Publisher, 88–103.

Burby, R. J. (2001). Flood insurance and floodplain management: the US experience. *Global Environmental Change Part B: Environmental Hazards, 3*(3), 111–122.

Burian, S. J. and Shepherd, J. M. (2005). Effect of urbanization on the diurnal rainfall pattern in Houston. *Hydrological Processes, 19*(5), 1089–1103.

Burnham, M. and Ma, Z. (2017). Climate change adaptation: factors influencing Chinese small-holder farmers' perceived self-efficacy and adaptation intent. *Regional Environmental Change, 17*(1), 171–186.

Burton, I. (1993). *The environment as hazard*: Guilford Press.

Bushley, B. R. (2014). REDD+ policy making in Nepal: toward state-centric, polycentric, or market-oriented governance? *Ecology and Society, 19*(3).

Calderon-Aguilera, L. E., Rivera-Monroy, V. H., Porter-Bolland, L., Martínez-Yrízar, A., Ladah, L. B., Martínez-Ramos, M., ... and Pérez-Salicrup, D. R. (2012). An assessment of natural and human disturbance effects on Mexican ecosystems: current trends and research gaps. *Biodiversity and Conservation, 21*(3), 589–617.

Cannaby, H., Fach, B. A., Arkin, S. S. and Salihoglu, B. (2015). Climatic controls on biophysical interactions in the Black Sea under present day conditions and a potential future (A1B) climate scenario. *Journal of Marine Systems, 141*, 149–166.

Carmin, J., Roberts, D. and Anguelovski, I. (2009). *Planning climate resilient cities: early lessons from early adapters.* Paper presented at the Proceedings of the World Bank Fifth Urban Research Symposium on Cities and Climate Change: Responding to an Urgent Agenda, Marseille, France.

Carpenter, S. R. and Gunderson, L. H. (2001). Coping with Collapse: Ecological and Social Dynamics in Ecosystem Management: Like flight simulators that train would-be aviators, simple models can be used to evoke people's adaptive, forward-thinking behavior, aimed in this instance at sustainability of human-natural systems. *AIBS Bulletin, 51*(6), 451–457.

Carpenter, S. R., Westley, F. and Turner, M. G. (2005). Surrogates for resilience of social-ecological systems. *Ecosystems, 8*(8), 941–944.

Cash, D. W., Clark, W. C., Alcock, F., Dickson, N. M., Eckley, N., Guston, D. H., ... Mitchell, R. B. (2003). Knowledge systems for sustainable development. *Proceedings of the national academy of sciences, 100*(14), 8086–8091.

Cattoën, C., McMillan, H. and Moore, S. (2016). Coupling a high-resolution weather model with a hydrological model for flood forecasting in New Zealand. *Journal of Hydrology, 55*(1), 1–23.

Cembrano, G., Quevedo, J., Salamero, M., Puig, V., Figueras, J. and Martí, J. (2004). Optimal control of urban drainage systems. A case study. *Control engineering practice, 12*(1), 1–9.

Chamlee-Wright, E. and Storr, V. H. (2009). "There's no place like New Orleans": Sense of place and community recovery in the Ninth Ward after Hurricane Katrina. *Journal of Urban Affairs, 31*(5), 615–634.

Chandra, A., Acosta, J., Meredith, L. S., Sanches, K., Stern, S., Uscher-Pines, L., Williams, M. and Yeung, D. (2010). *Building community resilience to disasters: A way forward to enhance national health security.* Part of working paper for Rand health.

Chao, Y. L. and Lam, S. P. (2011). Measuring responsible environmental behavior: Self-reported and other-reported measures and their differences in testing a behavioral model. *Environment and Behavior, 43*(1), 53–71.

Chapin, F. S., Kofinas, G. P., Folke, C. and Chapin, M. C. (2009). *Principles of ecosystem stewardship: resilience-based natural resource management in a changing world*: Springer Science and Business Media.

Che, W., Zhao, Y., Li, J., Wang, W., Wang, J., Wang, S. and Gong, Y. (2015). Exploreation of Sponge City Development Technical Guide: Basic Concepts and Comprehensive Goals. *China Water and Wastewater, 31*(8), 1–5.

Cheema, G. S. (1993). The challenge of urban management: Some issues. In *Urban management: Policies and innovations in developing countries*, 7.

Chen, L.-H., Yang, J.-L. and Barnes, M. D. (2014). Effect of natural disturbance on migrations and population dynamics of a catadromous species, Eriocheir japonicus, along the Lo-Mei stream in northern Taiwan. *Hydrobiologia, 726*(1), 179–194.

Chen, Y. and Yu, K. (2015). Thought of ancient "sponge city": experiences of applying water-adaptive landscape. *China Water Resources,* (17), 19–22.

Christmann, S. and Aw-Hassan, A. A. (2015). A participatory method to enhance the collective ability to adapt to rapid glacier loss: the case of mountain communities in Tajikistan. *Climatic Change, 133*(2), 267–282.

Clark, A. M., Friedman, E. J. and Hochstetler, K. (1998). The sovereign limits of global civil society: a comparison of NGO participation in UN world conferences on the environment, human rights, and women. *World politics, 51*(1), 1–35.

Clarke, S. E. (2017). Local place-based collaborative governance: Comparing state-centric and society-centered models. *Urban Affairs Review, 53*(3), 578–602.

Cohen, O., Bolotin, A., Lahad, M., Goldberg, A. and Aharonson-Daniel, L. (2016). Increasing sensitivity of results by using quantile regression analysis for exploring community resilience. *Ecological indicators, 66*, 497–502.

Coleman, J. S. (1986). Social theory, social research, and a theory of action. *American journal of Sociology, 91*(6), 1309–1335.

Colussi, M. (1999). The Community Resilience Manual - A New Resource Will Link Rural Revitalization To CED Best Practice. *Making Waves, 10*(4), 10–14.

Cosens, B., Gunderson, L., Arnold, A., Benson, M., Craig, R., DeCaro, D., Schlager, E. (2013). *Social-ecological system resilience, climate change, and adaptive water governance.* Project of the NSF funded National Socio-Environmental Synthesis Center, University of Maryland, Baltimore, Maryland, USA.

Cote, M. and Nightingale, A. J. (2012). Resilience thinking meets social theory Situating social change in socio-ecological systems (SES) research. *Progress in Human Geography, 36*(4), 475–489.

Cox, R. S. and Perry, K. M. E. (2011). Like a fish out of water: Reconsidering disaster recovery and the role of place and social capital in community disaster resilience. *American journal of community psychology, 48*(3–4), 395–411.

Cranton, P. (1994). *Understanding and Promoting Transformative Learning: A Guide for Educators of Adults. Jossey-Bass Higher and Adult Education Series*: ERIC.

Cranton, P. (1996). Types of group learning. *New Directions for Adult and Continuing Education, 1996*(71), 25–32.

Crijns, H., Cauberghe, V. and Hudders, L. (2017). Terrorism threat in Belgium: The resilience of Belgian citizens and the protection of governmental reputation by means of communication. *Public Relations Review, 43*(1), 219–234.

Cumming, G., Cumming, D. H. and Redman, C. (2006). Scale mismatches in social-ecological systems: causes, consequences, and solutions. *Ecology and Society, 11*(1).

Curran, S. R. and Saguy, A. C. (2001). Migration and cultural change: a role for gender and social networks? *Journal of International Women's Studies,* 2(3), 54–77.

Cutter, S. L. (2012). *Hazards vulnerability and environmental justice*: Routledge.

Cutter, S. L., Barnes, L., Berry, M., Burton, C., Evans, E., Tate, E. and Webb, J. (2008). A place-based model for understanding community resilience to natural disasters. *Global Environmental Change, 18*(4), 598–606.

Cutter, S. L. and Finch, C. (2008). Temporal and spatial changes in social vulnerability to natural hazards. *Proceedings of the national academy of sciences, 105*(7), 2301–2306.

Dale, V. H. and Beyeler, S. C. (2001). Challenges in the development and use of ecological indicators. *Ecological indicators, 1*(1), 3–10.

David, J. Y., Shin, H. C., Pérez, I., Anderies, J. M. and Janssen, M. A. (2016). Learning for resilience-based management: Generating hypotheses from a behavioral study. *Global Environmental Change, 37*, 69–78.

Davidson-Hunt, I. and Berkes, F. (2003). Learning as you journey: Anishinaabe perception of social-ecological environments and adaptive learning. *Conservation ecology, 8*(1).

Davoudi, S. (2012). Resilience: A Bridging Concept or a Dead End? *Planning Theory & Practice, 13*(2), 299–307.

de Loyola Hummell, B. M., Cutter, S. L. and Emrich, C. T. (2016). Social vulnerability to natural hazards in Brazil. *International Journal of Disaster Risk Science, 7*(2), 111–122.

Desaraju, V. R. and Michael, N. (2014). Hierarchical adaptive planning in environments with uncertain, spatially-varying disturbance forces. *Robotics and Automation (ICRA), 2014 IEEE International Conference*, 2014: 5171–5176.

Dhar, O. and Nandargi, S. (2003). *Hydrometeorological aspects of floods in India Flood Problem and Management in South Asia*: Springer.

Dixon, J. and Dupuis, A. (2003). *'Gatedness' and governance: residential intensification in Auckland, New Zealand*. Working paper at the Department of Urban Studies, University of Glasgow.

Dixon, P. G. and Mote, T. L. (2003). Patterns and causes of Atlanta's urban heat island-initiated precipitation. *Journal of Applied Meteorology, 42*(9), 1273–1284.

D'Odorico, P., Laio, F. and Ridolfi, L. (2010). Does globalization of water reduce societal resilience to drought? *Geophysical Research Letters, 37*(13).

Dolan, A. H. and Walker, I. (2006). Understanding vulnerability of coastal communities to climate change related risks. *Journal of Coastal Research, SI*(39), 1316–1323.

Dunlap, R. E., William R. and Catton J. (1979). Environmental sociology. *Annual Review of Sociology, 5*(1), 243–273.

Dunning, C. (1999). Post-intervention strategies to reduce police trauma: a paradigm shift. In *Police trauma: Psychological aftermath of civilian combat*, 269–289.

Ebersöhn, L. (2017). A resilience, health and well-being lens for education and poverty. *South African Journal of Education, 37*(1), 1–9.

Ellis, J. B. (2013). Sustainable surface water management and green infrastructure in UK urban catchment planning. *Journal of Environmental Planning and Management, 56*(1), 24–41.

Ensor, J., Park, S. E., Hoddy, E. and Ratner, B. (2015). A rights-based perspective on adaptive capacity. *Global Environmental Change, 31*, 38–49.

Etheredge, L. S. and Short, J. (1983). Thinking about government learning. *Journal of management studies, 20*(1), 41–58.

Eves, C. (2007). Planned residential community developments: do they add value? *Property Management, 25*(2), 164–179.

Fekete, A., Hufschmidt, G. and Kruse, S. (2014). Benefits and challenges of resilience and vulnerability for disaster risk management. *International Journal of Disaster Risk Science, 5*(1), 3–20.

Few, R. (2003). Flooding, vulnerability and coping strategies: local responses to a global threat. *Progress in Development Studies, 3*(1), 43–58.

Fewtrell, L. and Kay, D. (2008). An attempt to quantify the health impacts of flooding in the UK using an urban case study. *Public Health, 122*(5), 446–451.

Feyen, L., Barredo, J. I. and Dankers, R. (2009). *Implications of global warming and urban land use change on flooding in Europe*: CRC Press/Balkema, Leiden, The Netherlands.

Folke, C. (2006). Resilience: the emergence of a perspective for social-ecological systems analyses. *Global Environmental Change, 16*(3), 253–267.
Folke, C., Carpenter, S., Elmqvist, T., Gunderson, L., Holling, C. S. and Walker, B. (2002). Resilience and sustainable development: building adaptive capacity in a world of transformations. *AMBIO: A journal of the human environment, 31*(5), 437–440.
Folke, C., Carpenter, S. R., Walker, B., Scheffer, M., Chapin, T. and Rockstrom, J. (2010). Resilience thinking: integrating resilience, adaptability and transformability. *Ecology and society, 15*(4).
Folke, C., Colding, J. and Berkes, F. (2003). Synthesis: building resilience and adaptive capacity in social-ecological systems. *Navigating social-ecological systems: Building resilience for complexity and change, 9*(1), 352–387.
Folke, C., Hahn, T., Olsson, P. and Norberg, J. (2005). Adaptive governance of social-ecological systems. *Annu. Rev. Environ. Resour., 30*, 441–473.
Forbes, B. C., Stammler, F., Kumpula, T., Meschtyb, N., Pajunen, A. and Kaarlejärvi, E. (2009). High resilience in the Yamal-Nenets social-ecological system, West Siberian Arctic, Russia. *Proceedings of the national academy of sciences, 106*(52), 22041–22048.
Frazier, T. G., Thompson, C. M., Dezzani, R. J. and Butsick, D. (2013). Spatial and temporal quantification of resilience at the community scale. *Applied Geography, 42*, 95–107.
Friend, R. and Moench, M. (2013). What is the purpose of urban climate resilience? Implications for addressing poverty and vulnerability. *Urban Climate, 6*, 98–113.
Gaillard, J. C. (2007). Resilience of traditional societies in facing natural hazards. *Disaster prevention and management: An international journal, 16*(4), 522–544.
Gaillard, J. C. (2010). Vulnerability, capacity and resilience: perspectives for climate and development policy. *Journal of International Development, 22*(2), 218–232.
Galatzer-Levy, I. R., Steenkamp, M. M., Brown, A. D., Qian, M., Inslicht, S., Henn-Haase, C., ... Marmar, C. R. (2014). Cortisol response to an experimental stress paradigm prospectively predicts long-term distress and resilience trajectories in response to active police service. *Journal of Psychiatric Research, 56*, 36–42.
Gallopín, G. (2003). *A systems approach to sustainability and sustainable development* (Vol. 64): United Nations Publications.
Gionet, N. J. and Godin, G. (1989). Self-reported exercise behavior of employees: a validity study. *Journal of Occupational and Environmental Medicine, 31*(12), 969–973.
Glavovic, B., Scheyvens, R. and Overton, J. (2002). *Waves of adversity, layers of resilience: exploring the sustainable livelihoods approach.* Paper presented at the 3rd Biennial Conference of the International Development Network, New Zealand: Institute of Development Studies, Massey University.
Glendon, A. I., Clarke, S. and McKenna, E. (2016). *Human safety and risk management*: Crc Press.
Godschalk, D. R. (2003). Urban hazard mitigation: creating resilient cities. *Natural hazards review, 4*(3), 136–143.
Goldman, M. J. and Riosmena, F. (2013). Adaptive capacity in Tanzanian Maasailand: Changing strategies to cope with drought in fragmented landscapes. *Global Environmental Change, 23*(3), 588–597.
Goldstein, B. E. (2012). *Collaborative resilience: Moving through crisis to opportunity*: MIT Press.
González, J. A., Montes, C., Rodríguez, J. and Tapia, W. (2008). Rethinking the Galapagos Islands as a complex social-ecological system: implications for conservation and management. *Ecology and Society, 13*(2), 13.
Grimsey, D. and Lewis, M. K. (2002). Evaluating the risks of public private partnerships for infrastructure projects. *International Journal of Project Management, 20*(2), 107–118.
Guiver, J. W. (2013). The Role of Social Networks in providing Resilience in Travel Disruption. *Mobilities.*
Gunderson, L. H. (1999). Resilience, flexibility and adaptive management - antidotes for spurious certitude? *Conservation ecology, 3*(1).
Gunderson, L. H. (2000). Ecological resilience - in theory and application. *Annual review of ecology and systematics, 31*(1), 425–439.

Gunderson, L. H. (2010). Ecological and human community resilience in response to natural disasters. *Ecology and Society, 15*(2).

Guo, X., Fu, D., Guo, X. and Zhang, C. (2014). A case study of aerosol impacts on summer convective clouds and precipitation over northern China. *Atmospheric research, 142*, 142–157.

Guo, Y., Li, Y. and Ito, N. (2014). Exploring the predicted effect of social networking site use on perceived social capital and psychological well-being of Chinese international students in Japan. *Cyberpsychology, Behavior, and Social Networking, 17*(1), 52–58.

Gupta, K. (2007). Urban flood resilience planning and management and lessons for the future: a case study of Mumbai, India. *Urban Water Journal, 4*(3), 183–194.

Haimes, Y. Y., Crowther, K. and Horowitz, B. M. (2008). Homeland security preparedness: Balancing protection with resilience in emergent systems. *Systems Engineering, 11*(4), 287–308.

Hajat, S., Armstrong, B. G., Gouveia, N. and Wilkinson, P. (2005). Mortality displacement of heat-related deaths: a comparison of Delhi, Sao Paulo, and London. *Epidemiology, 16*(5), 613–620.

Han, L., Xu, Y., Lei, C., Yang, L., Deng, X., Hu, C. and Xu, G. (2016). Degrading river network due to urbanization in Yangtze River Delta. *Journal of Geographical Sciences, 26*(6), 694–706.

Han, L., Xu, Y., Shao, Y. and Shi, Y. (2013). Effect of urbanization on the stream structure and connectivity - A case study in the mid-lower reaches of the Qinhuai River. *Journal of Lake Sciences, 3*, 005.

Hao, S., Li, X. and Liu, J. (2014). Exploration and research on stormwater detention tank for reforming old urban storm sewer system. *China Water & Wastewater, 30*(16), 12–16.

Harbor, J. M. (1994). A practical method for estimating the impact of land-use change on surface runoff, groundwater recharge and wetland hydrology. *Journal of the American Planning Association, 60*(1), 95–108.

Harper, C. (2015). Organizations: Structures, processes and outcomes: *Routledge.*

He, B. (2003). The theory and practice of Chinese grassroots governance: Five models. *Japanese Journal of Political Science, 4*(2), 293–314.

Henly-Shepard, S., Gray, S. A. and Cox, L. J. (2015). The use of participatory modeling to promote social learning and facilitate community disaster planning. *Environmental Science and Policy, 45*, 109–122.

Hess, J. J., McDowell, J. Z. and Luber, G. (2012). Integrating climate change adaptation into public health practice: using adaptive management to increase adaptive capacity and build resilience. *Environmental Health Perspectives, 120*(2).

Hill, M. and Engle, N. L. (2013). Adaptive capacity: tensions across scales. *Environmental Policy and Governance, 23*(3), 177–192.

Hirokawa, K. H. (2014). *Environmental Law and Contrasting Ideas of Nature: A Constructivist Approach*: Cambridge University Press.

Holling, C. S. (1973). Resilience and stability of ecological systems. *Annual review of ecology and systematics, 4*(1), 1–23.

Holling, C. S. (1978). Adaptive environmental assessment and management. *Adaptive environmental assessment and management*: John Wiley & Sons.

Holling, C. S. (1986). The resilience of terrestrial ecosystems: local surprise and global change. In *Sustainable development of the biosphere*, 292–317.

Holling, C. S. (1996). Engineering resilience versus ecological resilience. In *Engineering within ecological constraints*, 31–44.

Holling, C. S. (2001). Understanding the complexity of economic, ecological, and social systems. *Ecosystems, 4*(5), 390–405.

Holling, C. S. and Gunderson, L. H. (2002). Resilience and adaptive cycles. *In: Panarchy: Understanding Transformations in Human and Natural Systems, 25–62.*

Holling, C. S., Gunderson, L. H. and Light, S. (1995). *Barriers and Bridges to the Renewal of Ecosystems*: Columbia University Press, New York.

Hollis, G. E. (1975). The effect of urbanization on floods of different recurrence interval. *Water Resources Research, 11*(3), 431–435.

Hollon, S. D. and Kendall, P. C. (1980). Cognitive self-statements in depression: Development of an automatic thoughts questionnaire. *Cognitive therapy and research, 4*(4), 383–395.

Hong, J. S. and Espelage, D. L. (2012). A review of research on bullying and peer victimization in school: An ecological system analysis. *Aggression and violent behavior, 17*(4), 311–322.

Horton, T. V. and Wallander, J. L. (2001). Hope and social support as resilience factors against psychological distress of mothers who care for children with chronic physical conditions. *Rehabilitation Psychology, 46*(4), 382.

Hou, L. (2015). Reflections on City Waterlogging Disaster Emergency Management and Its Countermeasures. *Public Administration & Law,* (1), 5–9.

Hughes, T. P., Bellwood, D. R., Folke, C., Steneck, R. S. and Wilson, J. (2005). New paradigms for supporting the resilience of marine ecosystems. *Trends in ecology & evolution, 20*(7), 380–386.

Huong, H. T. L. and Pathirana, A. (2013). Urbanization and climate change impacts on future urban flooding in Can Tho city, Vietnam. *Hydrology and Earth System Sciences, 17*(1), 379–394.

Ikeda, S., Sato, T. and Fukuzono, T. (2008). Towards an integrated management framework for emerging disaster risks in Japan. *Natural Hazards, 44*(2), 267–280.

IPCC. (2007). Climate change 2007: Impacts, adaptation and vulnerability. *Genebra, Suíça.*

Islam, R. and Walkerden, G. (2015). How do links between households and NGOs promote disaster resilience and recovery?: A case study of linking social networks on the Bangladeshi coast. *Natural Hazards, 78*(3), 1707–1727.

Ison, R., Röling, N. and Watson, D. (2007). Challenges to science and society in the sustainable management and use of water: investigating the role of social learning. *Environmental Science & Policy, 10*(6), 499–511.

Ito, J., Bao, Z. and Ni, J. (2016). Land rental development via institutional innovation in rural Jiangsu, China. *Food policy, 59*, 1–11.

Izadkhah, Y. O. and Hosseini, M. (2005). Towards resilient communities in developing countries through education of children for disaster preparedness. *International journal of emergency management, 2*(3), 138–148.

Izzati, T. (2016). An Initial Study Of The Air Pollution Through Rainwater In An Industrial Area Of Cikarang, West Java, Indonesia (A Case Study). *Science International (Lahore), 28*(4), 3695–3697.

Jabareen, Y. (2013). Planning the resilient city: Concepts and strategies for coping with climate change and environmental risk. *Cities, 31*, 220–229.

Jabeen, H., Johnson, C. and Allen, A. (2010). Built-in resilience: learning from grassroots coping strategies for climate variability. *Environment and Urbanization, 22*(2), 415–431.

Ji, X., Xu, Y., Han, L. and Yang, L. (2014). Impacts of urbanization on river system structure: a case study on Qinhuai River Basin, Yangtze River Delta. *Water Science and Technology, 70*(4), 671–677.

Jiang, F., Wei, Q., Yao, S., Wang, C. and Qu, X. (2013). Key theory and technical analysis on mine pressure bumping prevention and control. *Coal Science and Technology, 6*, 6–9.

Joseph, J. (2013). Resilience as embedded neoliberalism: a governmentality approach. *Resilience, 1*(1), 38–52.

Jung, M., Reichstein, M., Margolis, H. A., Cescatti, A., Richardson, A. D., Arain, M. A. and Gianelle, D. (2011). Global patterns of land-atmosphere fluxes of carbon dioxide, latent heat, and sensible heat derived from eddy covariance, satellite, and meteorological observations. *Journal of Geophysical Research: Biogeosciences, 116*(G3).

Kalnay, E. and Cai, M. (2003). Impact of urbanization and land-use change on climate. *Nature, 423*(6939), 528–531.

Kang, D. (2014). Discussion on problems and countermeasures in planning and design of urban drainage and waterlogging control systems. *China Water & Wastewater, 30*(20), 51–53.

Keck, M. and Sakdapolrak, P. (2013). What is social resilience? Lessons learned and ways forward. *Erdkunde, 67*(1), 5–19.

Keim, M. E. (2008). Building human resilience: the role of public health preparedness and response as an adaptation to climate change. *American journal of preventive medicine, 35*(5), 508–516.

Kendra, J. M. and Wachtendorf, T. (2003). Elements of resilience after the world trade center disaster: reconstituting New York City's Emergency Operations Centre. *Disasters, 27*(1), 37–53.

Khailani, D. K. and Perera, R. (2013). Mainstreaming disaster resilience attributes in local development plans for the adaptation to climate change induced flooding: A study based on the local plan of Shah Alam City, Malaysia. *Land Use Policy, 30*(1), 615–627.

Knighton, J., Lennon, E., Bastidas, L. and White, E. (2016). Stormwater detention system parameter sensitivity and uncertainty analysis using SWMM. *Journal of Hydrologic Engineering, 21*(8).

Konrad, C. P. (2003). *Effects of Urban Development on Floods*: United States Geological Survey Fact Sheet 076–03, Tacoma, Washington.

Koster, G. (2013). *Mapping runoff and erosion to reduce urban flooding and sediment flow towards sea: A case study on the Playa catchment, Bonaire.* Doctoral dissertation, Dissertation Calgary: Wageningen University.

Lam, S. P. and Cheng, S. I. (2002). Cross-informant agreement in reports of environmental behavior and the effect of cross-questioning on report accuracy. *Environment and Behavior, 34*(4), 508–520.

Le Goix, R. (2006). Gated communities as predators of public resources: the outcomes of fading boundaries between private management and public authorities in southern California in Glasze G Webster Cj and Frantz K eds. In *Private Cities: Global and local perspectives Routledge, London,* 76–91.

Legaspi, R., Maruyama, H., Nararatwong, R. and Okada, H. (2014). *Perception-based resilience: Accounting for the impact of human perception on resilience thinking.* Paper presented at the Big Data and Cloud Computing (BdCloud), 2014 IEEE Fourth International Conference.

Lei, Y., Yue, Y., Zhou, H. and Yin, W. (2014). Rethinking the relationships of vulnerability, resilience, and adaptation from a disaster risk perspective. *Natural Hazards, 70*(1), 609–627.

Leichenko, R. (2011). Climate change and urban resilience. *Current opinion in environmental sustainability, 3*(3), 164–168.

Leyden, K. M. (2003). Social capital and the built environment: the importance of walkable neighborhoods. *American journal of public health, 93*(9), 1546–1551.

Leykin, D., Lahad, M., Cohen, O., Goldberg, A. and Aharonson-Daniel, L. (2013). Conjoint community resiliency assessment measure-28/10 items (CCRAM28 and CCRAM10): A self-report tool for assessing community resilience. *American journal of community psychology, 52*(3–4), 313–323.

Li, M., Li, L. and Lan, H. (2014). Analysis of the Characteristics of Short Duration Heavy Precipitation Based on Data From High-density Automatic Weather Stations. *Science Technology and Engineering,* (18), 143–148.

Li, S., Zhu, Y. and Li, L. (2012). Neighborhood type, gatedness, and residential experiences in Chinese cities: A study of Guangzhou. *Urban Geography, 33*(2), 237–255.

Li, Z., Che, W. and Zhao, Y. (2015). Analysis of major drainage system of urban road in ancient china and its modern enlightenment. *China Water & Wastewater, 31*(10), 1–7.

Liao, K. H. (2012). A Theory on Urban Resilience to Floods - A Basis for Alternative Planning Practices. *Ecology and society, 17*(4), 48.

Lin, X. and Yu, S. (2005). Interdecadal changes of temperature in the Beijing region and its heat island effect. *Chinese Journal of Geophysics, 48*(1), 47–54.

Liu, B., Chen, J., Lu, W., Chen, X. and Lian, Y. (2016). Spatiotemporal characteristics of precipitation changes in the Pearl River Basin, China. *Theoretical and applied climatology, 123*(3–4), 537–550.

Liu, H., Liu, S., Zhu, J., Yin, Y. and Li, Y. (2014). Problems and policy-making for dealing urban flooding [城市内涝症结探讨及政策建议]. *Theory discussion, 24*(2): 39–41.

Liu, W., Ji, C., Zhong, J., Jiang, X. and Zheng, Z. (2007). Temporal characteristics of the Beijing urban heat island. *Theoretical and Applied Climatology, 87*(1), 213–221.

Liu, Z. (2015). *Institutional change in urbanizing China: a case study of rural shareholding cooperatives in Guangzhou.* Doctoral dissertation, Dissertation Calgary: The University of Hong Kong.

Lo, A. Y., Xu, B., Chan, F. K. and Su, R. (2015). Social capital and community preparation for urban flooding in China. *Applied Geography, 64*, 1–11.

Loo, B. P., Lam, W. W., Mahendran, R. and Katagiri, K. (2017). How Is the Neighborhood Environment Related to the Health of Seniors Living in Hong Kong, Singapore, and Tokyo? Some Insights for Promoting Aging in Place. *Annals of the American Association of Geographers, 107*(4), 812–828.

Lorenz, D. F. (2013). The diversity of resilience: contributions from a social science perspective. *Natural Hazards, 67*(1), 7–24.

Lyon, T. P. and Maxwell, J. W. (2008). Corporate social responsibility and the environment: A theoretical perspective. *Review of environmental economics and policy, 2*(2), 240–260.

Maarleveld, M. and Dabgbégnon, C. (1999). Managing natural resources: A social learning perspective. *Agriculture and human values, 16*(3), 267–280.

Maclean, K., Ross, H., Cuthill, M., & Witt, B. (2017). Converging disciplinary understandings of social aspects of resilience. *Journal of environmental planning and management, 60*(3), 519–537.

Marks, G., Hooghe, L. and Blank, K. (1996). European integration from the 1980s: State-centric v. multi-level governance. *Journal of Common Market Studies, 34*(3), 341–378.

Maguire, B. and Cartwright, S. (2008). Assessing a community's capacity to manage change: A resilience approach to social assessment. *Canberra: Bureau of Rural Sciences.*

Maguire, B. and Hagan, P. (2007). Disasters and communities: understanding social resilience. *Australian Journal of Emergency Management, 22*(2), 16–20.

Mallak, L. (1998). Putting organizational resilience to work. *Industrial Management (Norcross, Georgia), 40*(6), 8–13.

Manyena, S. B. (2006). The concept of resilience revisited. *Disasters, 30*(4), 434–450.

Martin, R. (2012). Regional economic resilience, hysteresis and recessionary shocks. *Journal of Economic Geography, 12*(1), 1–32.

Mason, Y., Ammann, A. A., Ulrich, A. and Sigg, L. (1999). Behavior of heavy metals, nutrients, and major components during roof runoff infiltration. *Environmental science & technology, 33*(10), 1588–1597.

Masten, A. S. (2015). *Ordinary magic: Resilience in development*: Guilford Publications.

Maybery, D., Pope, R., Hodgins, G., Hitchenor, Y. and Shepherd, A. (2009). Resilience and well-being of small inland communities: Community assets as key determinants. *Rural Society, 19*(4), 326–339.

McGill, R. (1998). Urban management in developing countries. *Cities, 15*(6), 463–471.

McGinnis, M. and Ostrom, E. (2014). Social-ecological system framework: initial changes and continuing challenges. *Ecology and Society, 19*(2).

McIlwaine, C. and Datta, K. (2014). Sustaining a global city at work: resilient geographies of a migrant division of labour. In *Sustainable London?: The future of a global city*, 111–128.

McIntosh, R. J. (2000). Social memory in Mande. In *The way the wind blows: climate, history, and human action*, 141–180.

McPhearson, T., Hamstead, Z. A. and Kremer, P. (2014). Urban ecosystem services for resilience planning and management in New York City. *Ambio, 43*(4), 502–515.

Mezirow, J. (1991). *Transformative dimensions of adult learning*: ERIC.

Mezirow, J. (2000). *Learning as Transformation: Critical Perspectives on a Theory in Progress. The Jossey-Bass Higher and Adult Education Series*: ERIC.

Mezirow, J. (2006). An overview on transformative learning. In *Lifelong learning: Concepts and contexts*, 90–106.

Miao, S., Chen, F., Li, Q. and Fan, S. (2011). Impacts of urban processes and urbanization on summer precipitation: A case study of heavy rainfall in Beijing on 1 August 2006. *Journal of Applied Meteorology and Climatology, 50*(4), 806–825.

Michie, S., Fixsen, D., Grimshaw, J. M. and Eccles, M. P. (2009). Specifying and reporting complex behaviour change interventions: the need for a scientific method. *Implementation Science, 4*(1).

Mills, M. B. (1997). Contesting the margins of modernity: Women, migration, and consumption in Thailand. *American ethnologist, 24*(1), 37–61.

Mirza, M. M. Q. (2003). Climate change and extreme weather events: can developing countries adapt? *Climate policy, 3*(3), 233–248.

Montalto, F., Behr, C., Alfredo, K., Wolf, M., Arye, M. and Walsh, M. (2007). Rapid assessment of the cost-effectiveness of low impact development for CSO control. *Landscape and urban planning, 82*(3), 117–131.

Montano, D. E. and Kasprzyk, D. (2015). Theory of reasoned action, theory of planned behavior, and the integrated behavioral model. In *Health behavior: Theory, research and practice,* 95–125.

Montoya, L. (2003). Geo-data acquisition through mobile GIS and digital video: an urban disaster management perspective. *Environmental Modelling & Software, 18*(10), 869–876.

Moore, M. L. and Westley, F. (2011). Surmountable chasms: networks and social innovation for resilient systems. *Ecology and society, 16*(1).

Mostert, E., Pahl-Wostl, C., Rees, Y., Searle, B., Tàbara, D. and Tippett, J. (2007). Social learning in European river-basin management: barriers and fostering mechanisms from 10 river basins. *Ecology and Society, 12*(1).

Mote, T. L., Lacke, M. C. and Shepherd, J. M. (2007). Radar signatures of the urban effect on precipitation distribution: a case study for Atlanta, Georgia. *Geophysical Research Letters, 34*(20).

Naumann, T., Nikolowski, J. and Golz, S. (2009). *Synthetic depth-damage functions–a detailed tool for analysing flood resilience of building types.* Road map towards a flood resilient urban environment, Final conference of the COST action C 22 Urban Flood Management in cooperation with UNESCO-IHP, edited by: Pasche, E., Evelpidou, N., Zevenbergen, C., Ashley, R., and Garvin, S., Paris, 26–27 November 2009.

Newig, J. and Fritsch, O. (2009). Environmental governance: participatory, multi-level–and effective? *Environmental Policy and Governance, 19*(3), 197–214.

Newman, L. and Dale, A. (2005). Network structure, diversity, and proactive resilience building: a response to Tompkins and Adger. *Ecology and Society, 10*(1).

Nguyen, K. and James, H. (2013). Measuring household resilience to floods: a case study in the Vietnamese Mekong River Delta. *Ecology and Society, 18*(3).

Nicholls, R. J. and de la Vega-Leinert, A. C. (2008). Implications of sea-level rise for Europe's coasts: an introduction. *Journal of Coastal Research, 24*(2), 285–287.

Nickel, D., Schoenfelder, W., Medearis, D., Dolowitz, D. P., Keeley, M. and Shuster, W. (2014). German experience in managing stormwater with green infrastructure. *Journal of environmental planning and management, 57*(3), 403–423.

Nie, L., Lindholm, O., Lindholm, G. and Syversen, E. (2009). Impacts of climate change on urban drainage systems–a case study in Fredrikstad, Norway. *Urban Water Journal, 6*(4), 323–332.

Norio, O., Ye, T., Kajitani, Y., Shi, P. and Tatano, H. (2011). The 2011 eastern Japan great earthquake disaster: Overview and comments. *International Journal of Disaster Risk Science, 2*(1), 34–42.

Norris, F. H., Stevens, S. P., Pfefferbaum, B., Wyche, K. F. and Pfefferbaum, R. L. (2008). Community resilience as a metaphor, theory, set of capacities, and strategy for disaster readiness. *American journal of community psychology, 41*(1–2), 127–150.

O'Brien, G., O'Keefe, P., Rose, J. and Wisner, B. (2006). Climate change and disaster management. *Disasters, 30*(1), 64–80.

Obrist, B., Pfeiffer, C. and Henley, R. (2010). Multi-layered social resilience: A new approach in mitigation research. *Progress in Development Studies, 10*(4), 283–293.

Olsson, P., Folke, C. and Berkes, F. (2004). Adaptive comanagement for building resilience in social–ecological systems. *Environmental management, 34*(1), 75–90.

O'Rourke, T. D. (2007). Critical infrastructure, interdependencies, and resilience. *Bridge-Washington-National Academy of Engineering, 37*(1), 22.

Owa, F. W. (2014). Water pollution: sources, effects, control and management. *International Letters of Natural Sciences, 3.*

Pahl-Wostl, C. (2007). Transitions towards adaptive management of water facing climate and global change. *Water resources management, 21*(1), 49–62.

Pálsson, G. (2003). Human-environmental relations: orientalism, paternalism and communalism. In *Nature and society*. Routledge, 73–91.

Parsons, L. (2016). Environmental Risk and Contemporary Resilience Strategies in Rural Cambodia. In *The Handbook of Contemporary Cambodia*. Routledge, 146–156.

Paton, D. (2006). Disaster resilience: integrating individual, community, institutional and environmental perspectives. In *Disaster resilience: An integrated approach*. Charles C Thomas Publisher, 305–318.

Paton, D. and Johnston, D. (2017). *Disaster resilience: an integrated approach*: Charles C Thomas Publisher.

Paton, D., Smith, L. and Violanti, J. (2000). Disaster response: risk, vulnerability and resilience. *Disaster prevention and management: An international journal, 9*(3), 173–180.

Pattee, H. H. (2012). Evolving self-reference: matter, symbols, and semantic closure. In *LAWS, LANGUAGE and LIFE*. Springer, 211–226.

Peipoch, M., Brauns, M., Hauer, F. R., Weitere, M. and Valett, H. M. (2015). Ecological simplification: human influences on riverscape complexity. *BioScience, 65*(11), 1057–1065.

Pelling, M. (2003). *The vulnerability of cities: natural disasters and social resilience*: Earthscan.

Pelling, M., Sharpe, J., Pearson, L., Abeling, T., Gerger Swartling, A., Forrester, J. M. and Deeming, H. (2015). Social Learning and Resilience Building in the emBRACE framework.

Pendall, R., Foster, K. A. and Cowell, M. (2010). Resilience and regions: building understanding of the metaphor. *Cambridge Journal of Regions, Economy and Society, 3*(1), 71–84.

Penn, D. and Mysterud, I. (2007). Introduction: the evolutionary roots of our ecological crisis. *Evolutionary perspectives on environmental problems. New Brunswick (NJ): Aldine Transaction*, 1–8.

Perrings, C. (2006). Resilience and sustainable development. *Environment and Development Economics, 11*(04), 417–427.

Peterson, G. (2000). Political ecology and ecological resilience: An integration of human and ecological dynamics. *Ecological Economics, 35*(3), 323–336.

Plieninger, T. and Bieling, C. (2012). *Resilience and the cultural landscape: understanding and managing change in human-shaped environments*: Cambridge University Press.

Plough, A., Fielding, J. E., Chandra, A., Williams, M., Eisenman, D., Wells, K. B., ... and Magaña, A. (2013). Building community disaster resilience: perspectives from a large urban county department of public health. *American journal of public health, 103*(7), 1190–1197.

Plummer, R. and Armitage, D. (2010). Integrating perspectives on adaptive capacity and environmental governance. In *Adaptive capacity and environmental governance*. Springer, 1–19.

Podsakoff, P. M., MacKenzie, S. B., Lee, J.-Y. and Podsakoff, N. P. (2003). Common method biases in behavioral research: a critical review of the literature and recommended remedies. *Journal of applied psychology, 88*(5), 879.

Pow, C. P. and Kong, L. (2007). Marketing the Chinese dream home: Gated communities and representations of the good life in (post-) socialist Shanghai. *Urban Geography, 28*(2), 129–159.

Qin, H., Li, Z. and Fu, G. (2013). The effects of low impact development on urban flooding under different rainfall characteristics. *Journal of environmental management, 129*, 577–585.

Qiu, B. (2015a). Concept, pathway, and future development of Sponge City (LID). *Construction Science and Technology, 41*(3), 1–7.

Qiu, B. (2015b). Focus on Sponge City Development - Concept, pathway and future development of Sponge City (LID), *China Investigation & Design*, (7), 28–41.

Quarantelli, E. L. (1998). Major criteria for judging disaster planning and managing their applicability in developing societies. In *Preliminary Paper* (Vol. 268). US University of Delaware. Disaster Research Center (DRC).

Rakodi, C. (1991). Cities and people; towards a gender-aware urban policy process? *Public Administration and Development*, 11, 541–559.

Ramalingam, B. (2015). The Demand For, Feasibility and Scope of a Global Network of Southern NGOs in Disaster Resilience, Response and Recovery: ADESO.

Redman, C. and Kinzig, A. (2003). Resilience of past landscapes: resilience theory, society, and the longue durée. *Conservation ecology, 7*(1).

Reed, M. S. (2008). Stakeholder participation for environmental management: a literature review. *Biological conservation, 141*(10), 2417–2431.

Rigg, J. and Oven, K. (2015). Building liberal resilience? A critical review from developing rural Asia. *Global Environmental Change,* 32, 175–186.

Rogers, D. P. and Tsirkunov, V. V. (2013). Weather and Climate Resilience. *Direction in Development.*

Rose, A. (2004). Defining and measuring economic resilience to disasters. *Disaster prevention and management: An international journal, 13*(4), 307–314.

Sakieh, Y., Amiri, B. J., Danekar, A., Feghhi, J. and Dezhkam, S. (2015). Scenario-based evaluation of urban development sustainability: an integrative modeling approach to compromise between urbanization suitability index and landscape pattern. *Environment, Development and Sustainability, 17*(6), 1343–1365.

Saldaña-Zorrilla, S. O. (2008). Stakeholders' views in reducing rural vulnerability to natural disasters in Southern Mexico: Hazard exposure and coping and adaptive capacity. *Global Environmental Change, 18*(4), 583–597.

Sampson, R. J. (2003). The neighborhood context of well-being. *Perspectives in biology and medicine, 46*(3), S53–S64.

Sartor, J. D. and Boyd, G. B. (1972). *Water pollution aspects of street surface contaminants* (Vol. 81). US Government Printing Office.

Shamsudin, Z. B., Ying, E. and Omar, A. (2016). The Safety Level Satisfaction towards Gated and Guarded Community in Malaysia. *International Review of Management and Marketing, 6* (7S).

Shao Y. and Xu J. (2015). Resilient cities: a new shift to urban crisis management. *Urban Planning Internatioanl*, 2: 1–3.

Sharpe, J. (2016). Understanding and unlocking transformative learning as a method for enabling behaviour change for adaptation and resilience to disaster threats. *International journal of disaster risk reduction, 17*, 213–219.

Shastri, H., Paul, S., Ghosh, S. and Karmakar, S. (2015). Impacts of urbanization on Indian summer monsoon rainfall extremes. *Journal of Geophysical Research: Atmospheres, 120*(2), 496–516.

Shaw, R., Kobayashi, K. and Kobayashi, M. (2004). Linking experience, education, perception and earthquake preparedness. *Disaster prevention and management: An international journal, 13*(1), 39–49.

Shen, W. B. (2013). Improvement techonologies for urban sewerage systems [城市排水管网系统改造技术]. *China Construction,* (12), 188–189.

Shepherd, J. M., Pierce, H. and Negri, A. J. (2002). Rainfall modification by major urban areas: Observations from spaceborne rain radar on the TRMM satellite. *Journal of Applied Meteorology, 41*(7), 689–701.

Shi, P., Yuan, Y., Zheng, J., Wang, J., Ge, Y. and Qiu, G. (2007). The effect of land use/cover change on surface runoff in Shenzhen region, China. *Catena, 69*(1), 31–35.

Siegrist, J., Anderson, D., Koran, J., Pribak, M., Shamsi, U.M.S. and White, D. (2016). Assessing SWMM 5 Hydrologic Parameter Benefits for Model Calibration. *Journal of Water Management Modeling, 25.*

Silver, A. and Grek-Martin, J. (2015). "Now we understand what community really means": Reconceptualizing the role of sense of place in the disaster recovery process. *Journal of environmental psychology, 42*, 32–41.

Simmie, J. and Martin, R. (2010). The economic resilience of regions: towards an evolutionary approach. *Cambridge journal of regions, economy and society, 3*(1), 27–43.

Simon, H. A. (1977). The organization of complex systems. In *Models of discovery.* Springer, 245–261.

Slocombe, D. S. (1993). Implementing ecosystem-based management. *BioScience, 43*(9), 612–622.

Smith, A. and Stirling, A. (2010). The politics of social-ecological resilience and sustainable socio-technical transitions. *Ecology and Society, 15*(1).

Smith, J. G., DuBois, B. and Krasny, M. E. (2016). Framing for resilience through social learning: impacts of environmental stewardship on youth in post-disturbance communities. *Sustainability Science, 11*(3), 441–453.

Smith, J. W., Anderson, D. H. and Moore, R. L. (2012). Social capital, place meanings, and perceived resilience to climate change. *Rural Sociology, 77*(3), 380–407.

Smith, N. R. (2014). Building resilient cities in China: the nexus between planning and science. *Town Planning Review, 85*(1), 127–142.

Soetanto, R., Mullins, A. and Achour, N. (2016). The perceptions of social responsibility for community resilience to flooding: the impact of past experience, age, gender and ethnicity. *Natural Hazards*, 1–22.

Soler, J., Cebolla, A., Feliu-Soler, A., Demarzo, M. M., Pascual, J. C., Baños, R. and García-Campayo, J. (2014). Relationship between meditative practice and self-reported mindfulness: the MINDSENS composite index. *PloS one, 9*(1), e86622.

Song, L., Tian, Y., Wu, L. and Zhang, H. (2008). On Comparison of Spatial Interpolation Methods of Daily Rainfall Data: A Case Study of Shenzhen. *Geo-information science, 5*, 004.

Srinivasan, V., Seto, K. C., Emerson, R. and Gorelick, S. M. (2013). The impact of urbanization on water vulnerability: a coupled human–environment system approach for Chennai, India. *Global Environmental Change, 23*(1), 229–239.

Sterling, S. (2010). Learning for resilience, or the resilient learner? Towards a necessary reconciliation in a paradigm of sustainable education. *Environmental Education Research, 16*(5–6), 511–528.

Steward, F. (2012). Transformative innovation policy to meet the challenge of climate change: sociotechnical networks aligned with consumption and end-use as new transition arenas for a low-carbon society or green economy. *Technology Analysis & Strategic Management, 24*(4), 331–343.

Stokols, D. (1978). Environmental psychology. *Annual review of psychology, 29*(1), 253–295.

Stokols, D. and Altman, I. (1987). *Handbook of environmental psychology* (Vol. 2): Wiley.

Stokols, D., Lejano, R. and Hipp, J. (2013). Enhancing the resilience of human–environment systems: A social ecological perspective. *Ecology and Society, 18*(1).

Stone-Jovicich, S., Goldstein, B., Brown, K., Plummer, R. and Olsson, P. (2018). Expanding the contribution of the social sciences to social-ecological resilience research. *Ecology and Society,* 23(1).

Sugimoto, M., Iemura, H. and Shaw, R. (2010). Tsunami height poles and disaster awareness: Memory, education and awareness of disaster on the reconstruction for resilient city in Banda Aceh, Indonesia. *Disaster prevention and management: An international journal, 19*(5), 527–540.

Sullivan, H. T. and Häkkinen, M. T. (2006). *Disaster preparedness for vulnerable populations: Determining effective strategies for communicating risk, warning, and response.* Present at the Third Annual Magrann Research Conference at Rutgers University.

Sun, A. L. (2011). *Risk assessment of rainstorm waterlogging based on scenario simulation.* Master dissertation, Dissertation Calgary: East China Normal University.

Suzuki, Y., Ogoshi, M., Yamagata, H., Ozaki, M. and Asano, T. (2002). *Large-area and on-site water reuse in Japan.* Paper present at the International Seminar on Sustainable Development of Water Resource Pluralizing Technology, held at Korean National Institute of Environmental Research.

Tanaka, K. (2005). The impact of disaster education on public preparation and mitigation for earthquakes: a cross-country comparison between Fukui, Japan and the San Francisco Bay Area, California, USA. *Applied Geography, 25*(3), 201–225.

Tanner, T., Mitchell, T., Polack, E. and Guenther, B. (2009). Urban governance for adaptation: assessing climate change resilience in ten Asian cities. *IDS Working Papers, 2009*(315), 01–47.

Terti, G., Ruin, I., Anquetin, S. and Gourley, J. J. (2015). Toward a coupled Hazard-Vulnerability Tool for Flash Flood Impacts Prediction. In *EGU General Assembly Conference Abstracts* (Vol. 17).

Tingsanchali, T. (2012). Urban flood disaster management. *Procedia engineering, 32*, 25–37.

Tran, P. and Shaw, R. (2007). Towards an integrated approach of disaster and environment management: A case study of Thua Thien Hue province, central Viet Nam. *Environmental Hazards, 7*(4), 271–282.

Tschakert, P. and Dietrich, K. (2010). Anticipatory learning for climate change adaptation and resilience. *Ecology and Society, 15*(2).

Tunstall, S. M., Johnson, C. L. and Penning-Rowsell, E. C. (2004). Flood hazard management in England and Wales: from land drainage to flood risk management. In *World Congress on Natural Disaster Mitigation*, 19–21.

Turner, B. A. (1994). Causes of disaster: sloppy management. *British Journal of management, 5*(3), 215–219.

Turner, M. D. (2014). Political ecology I: An alliance with resilience?. *Progress in Human Geography, 38*(4), 616–623.

Twigg, J. and Steiner, D. (2002). Mainstreaming disaster mitigation: challenges to organisational learning in NGOs. *Development in Practice, 12*(3–4), 473–479.

Ungar, M. (2012). Social ecologies and their contribution to resilience *The social ecology of resilience*: Springer.

United Nations (UN), *World Urbanization Prospects: The 2005 Revision* (New York: United Nations, 2006).

Van Wormer, K. (2017). *Human behavior and the social environment, micro level: Individuals and families*: Oxford University Press.

Vedung, E. (2017). *Public policy and program evaluation*: Routledge.

Veronesi, M., Chawla, F., Maurer, M. and Lienert, J. (2014). Climate change and the willingness to pay to reduce ecological and health risks from wastewater flooding in urban centers and the environment. *Ecological Economics, 98*, 1–10.

Vugrin, E. D. and Camphouse, R. C. (2011). Infrastructure resilience assessment through control design. *International journal of critical infrastructures, 7*(3), 243–260.

Walker, B., Carpenter, S., Anderies1b, J., Abel1b, N., Cumming, G., Janssen, M., . . . Pritchard, R. (2002). Resilience management in social-ecological systems: a working hypothesis for a participatory approach. *Conservation ecology, 6*(1), 14.

Walker, B., Gunderson, L., Kinzig, A., Folke, C., Carpenter, S. and Schultz, L. (2006). A handful of heuristics and some propositions for understanding resilience in social-ecological systems. *Ecology and society, 11*(1), 13.

Walker, B., Holling, C. S., Carpenter, S. and Kinzig, A. (2004). Resilience, adaptability and transformability in social–ecological systems. *Ecology and Society, 9*(2).

Walklate, S. and Mythen, G. (2014). *Contradictions of Terrorism: Security, Risk and Resilience*: Routledge.

Walsh, F. (2015). *Strengthening family resilience*: Guilford Publications.

Walters, C. J. and Holling, C. S. (1990). Large-scale management experiments and learning by doing. *Ecology, 71*(6), 2060–2068.

Wang, S., Ma, H. and Zhao, Y. (2014). Exploring the relationship between urbanization and the eco-environment—a case study of Beijing-Tianjin-Hebei region. *Ecological Indicators, 45*, 171–183.

Wang, S. H., Huang, S. L. and Budd, W. W. (2012). Resilience analysis of the interaction of between typhoons and land use change. *Landscape and urban planning, 106*(4), 303–315.

Welsh, M. (2014). Resilience and responsibility: governing uncertainty in a complex world. *The Geographical Journal, 180*(1), 15–26.

Wheater, H. S. (2006). Flood hazard and management: a UK perspective. *Philosophical Transactions of the Royal Society of London A: Mathematical, Physical and Engineering Sciences, 364*(1845), 2135–2145.

White, I. (2013). *Water and the city: Risk, resilience and planning for a sustainable future*: Routledge.

White, M. D. and Greer, K. A. (2006). The effects of watershed urbanization on the stream hydrology and riparian vegetation of Los Penasquitos Creek, California. *Landscape and Urban Planning, 74*(2), 125–138.

Wickes, R., Zahnow, R., Taylor, M. and Piquero, A. R. (2015). Neighborhood structure, social capital, and community resilience: longitudinal evidence from the 2011 Brisbane flood disaster. *Social Science Quarterly, 96*(2), 330–353.

Wieland, A. and Marcus Wallenburg, C. (2013). The influence of relational competencies on supply chain resilience: a relational view. *International Journal of Physical Distribution & Logistics Management, 43*(4), 300–320.

Wildavsky, A. B. (1988). *Searching for safety* (Vol. 10): Transaction publishers.

Willems, P., Olsson, J. and Arnbjerg-Nielsen, K. (2012). *Impacts of climate change on rainfall extremes and urban drainage systems*: IWA Publishing.

Wilson, S., Pearson, L., Kashima, Y., Lusher, D. and Pearson, C. (2013). Separating adaptive maintenance (resilience) and transformative capacity of social-ecological systems. *Ecology and Society, 18*(1).

Windle, G. (2011). What is resilience? A review and concept analysis. *Reviews in Clinical Gerontology, 21*(02), 152–169.

Wong, S. W., Tang, B. S. and van Horen, B. (2006). Strategic urban management in China: A case study of Guangzhou Development District. *Habitat International, 30*(3), 645–667.

Xiang, W. (2014). Doing real and permanent good in landscape and urban planning: Ecological wisdom for urban sustainability. *Landscape and Urban Planning, 121*, 65–69.

Xie, Y. (2013). Development of drainage planning in view of frequent urban waterlogging disasters. *City Planning Review*, (2), 45–50.

Xu, J. and Ou, L. (2014). Resilience and quality of life among Wenchuan earthquake survivors: the mediating role of social support. *Public Health, 128*(5), 430–437.

Yang, L., Smith, J. A., Wright, D. B., Baeck, M. L., Villarini, G., Tian, F. and Hu, H. (2013). Urbanization and climate change: an examination of nonstationarities in urban flooding. *Journal of Hydrometeorology, 14*(6), 1791–1809.

Ye, H., Fetzer, E. J., Behrangi, A., Wong, S., Lambrigtsen, B. H., Wang, C. Y. and Gamelin, B. L. (2016). Increasing daily precipitation intensity associated with warmer air temperatures over Northern Eurasia. *Journal of Climate, 29*(2), 623–636.

Ye, Y. M. (2008). Citi-PODAS Model: A Brave Trial from Traditional to Modern Urban Management in China. *China Soft Science,* (10), 52–64.

Yohe, G. and Tol, R. S. (2002). Indicators for social and economic coping capacity - moving toward a working definition of adaptive capacity. *Global Environmental Change, 12*(1), 25–40.

Youssef, K. W. F. (2015). *Gated-ness, Income Segregation and Neighbourhood Cohesion in Two Western Canadian Metropolises.* Doctoral dissertation, Dissertation Calgary: University of Calgary.

Yu, H., Huang, G. and Wu, C. (2012). Study on Urban Storm Water Modeling at Jinan City Using SWMM Model. *Journal of Water Resources Research, Research of Soil And Water Conservation, 1*, 433–439.

Yu, K, Li, D., Yuan, H., Fu, W., Qiao, Q. and Wang, S. (2015). "Sponge City": Theory And Practice. *City Planning Review, 39*(6), 26 –36.

Yucel, I., Onen, A., Yilmaz, K. K. and Gochis, D. J. (2015). Calibration and evaluation of a flood forecasting system: Utility of numerical weather prediction model, data assimilation and satellite-based rainfall. *Journal of Hydrology, 523*, 49–66.

Zahmatkesh, Z., Burian, S. J., Karamouz, M., Tavakol-Davani, H. and Goharian, E. (2014). Low-impact development practices to mitigate climate change effects on urban stormwater runoff: Case study of New York City. *Journal of Irrigation and Drainage Engineering, 141*(1).

Zamani, B. and Arefi, M. (2013). Iranian New Towns and their Urban Management Issues: A critical review of influential actors and factors. *Cities, 30*, 105–112.

Zhang, J. and Lou, C. (2014). Study on the Metropolis Waterlogging Prevention in the View of Urban Governance. *The Journal of Shanghai Administration Institute, 15*(4), 31–39.
Zhang, X., Cheung D. M., Sun, Y, and Tan, J. (2019). Political Decentralization and the Path Dependent Characteristics of the State Authoritarianism: An Integrated Conceptual Framework to Understand China's Territorial Fragmentation. *Eurasian Geography and Economics, 60*(5), 548–581.
Zhang, X., Shen, J., and Gao, X. (2021).Towards a Comprehensive Understanding of Intercity Cooperation in China's City-regionalization: A Comparative Study of Shenzhen-Hong Kong and Guangzhou-Foshan City Groups. *Land Use Policy,103*,105339.
Zhang, X., Li, Y., Zhou, C., Luan, X., and Yuan F. (2023). Rescaling of the Land Regime in the Making of City-regions: A Case Study of China's Pearl River Delta.*Urban Studies,60*(3), 483–500.
Zhang, X., Shen, J., Sun, Y., Zhou, C. and Yang, Y. (2023). Formation of City regions from the Bottom-up Initiatives: Investigating Coalitional Developmentalism in the Pearl River Delta. *Annals of the American Association of Geographers, 113*(3), 700–716.
Zhang, X., Zhou, Z., and Yang, Y. (2023). Stepwise Transition of Urban Entrepreneuralism in Cross-boundary Statecraft: A Qualitative Study of An Enclave Industrial Park in Zhejiang Province, China. *Transactions in Planning and Urban Research, 2*(1), 85–102.
Zhang, X., Chen, Y., Xu, Y., Yue, W., and Tang, C. (2023). Does regional cooperation constrain urban sprawl? Evidence from the Guangdong-Hong Kong-Macao Greater Bay Area. *Landscape and Urban Planning*, 235, 104742.
Zheng, Z., Wang, Z. and Gao, H. (2013). Characteristics of Extreme Precipitation Events in Summer and Its Effect on Urbanization in Beijing Area. *Neteorological Monthly, 39*(12), 1635–1641.
Zhu, L. (2013). Deal with urban problems before urban flooding management [治内涝先治城市病]. *Zhongzhou Construction,* (17), 44.
Ziska, L. H., Gebhard, D. E., Frenz, D. A., Faulkner, S., Singer, B. D. and Straka, J. G. (2003). Cities as harbingers of climate change: common ragweed, urbanization, and public health. *Journal of Allergy and Clinical Immunology, 111*(2), 290–295.

MIX
Papier aus verantwortungsvollen Quellen
Paper from responsible sources
FSC® C105338

If you have any concerns about our products,
you can contact us on
ProductSafety@springernature.com

In case Publisher is established outside the EU,
the EU authorized representative is:
Springer Nature Customer Service Center GmbH
Europaplatz 3, 69115 Heidelberg, Germany

Printed by Libri Plureos GmbH
in Hamburg, Germany